OXFORD STUDIES IN PHYSICS

GENERAL EDITORS

# THEORY OF LEPTON–HADRON PROCESSES AT HIGH ENERGIES

## PARTONS, SCALE INVARIANCE, AND LIGHT-CONE PHYSICS

BY

PROBIR ROY

*Tata Institute of Fundamental Research*
*Colaba, Bombay*



CLARENDON PRESS · OXFORD

1975

*Oxford University Press, Ely House, London W.1*
GLASGOW NEW YORK TORONTO MELBOURNE WELLINGTON
CAPE TOWN IBADAN NAIROBI DAR ES SALAAM LUSAKA ADDIS ABABA
DELHI BOMBAY CALCUTTA MADRAS KARACHI LAHORE DACCA
KUALA LUMPUR SINGAPORE TOKYO

ISBN 0 19 851452 2



PRINTED IN GREAT BRITAIN
BY J. W. ARROWSMITH LTD., BRISTOL, ENGLAND

TO

MANASHI

# PREFACE

THE recent interest in high energy lepton–hadron processes originates from two factors. First, sufficiently energetic lepton beams have become available for probing the structure of hadrons in collisions which involve currents with high virtual masses. Second, the measurement, at SLAC, of sizeable inelastic cross-sections for electrons imparting large momentum transfers to target nucleons has led to the discovery of the Bjorken scaling phenomenon in these data. It would not be an exaggeration to say that the SLAC experiments have been the major motive force behind the development of concepts such as partons, short-distance scale invariance, and leading free-field behaviour on the light-cone. Of course the theory of deep inelastic lepton–hadron reactions is admittedly still in a state of flux. Further accelerator experiments at Cornell, SLAC, NAL, and CERN and storage-ring experiments at Frascati, Orsay, DORIS, CEA, SPEAR, and Novosibirsk are being conducted and planned. These experiments promise a rich outflow of new data in future and correspondingly intense theoretical activity, leading to possible new breakthroughs. Nevertheless, certain broad, basic, clear-cut features have already emerged that merit attention.

The aim of this book is to provide physicists with a compact, reasonably complete, yet elementary introduction to the main streams of the theoretical developments mentioned above. I have tried to explain and treat in perspective key aspects of the subject in a self-contained manner. These aspects have historically evolved together and at present display a basic unity in approaching the problem of deep inelastic lepton–hadron processes. It is hoped that the book will prove useful to research students and high-energy experimentalists, as well as to the working theoretical physicist who is not necessarily a specialist in this field.

The Introduction, summarizing the salient kinematic and empirical features of the subject, is followed by three parts. In Part I the parton picture is considered. This model provides a penetrating insight into Bjorken scaling and moreover leads to a large number of testable results in broad conformity with present experiment. The theoretical basis of the parton picture is also discussed in terms of more fundamental dynamical considerations. Part II is devoted to the study of scale invariance. The notions of exact and partial invariance under dilations are first clarified; then the role of such invariance in the short-distance behaviour of products of hadronic currents, and hence in the asymptotic properties of inelastic lepton–hadron reactions, is discussed. An account of light-cone physics is given in Part III. The historical evolution of light-cone expansions is sketched first; this is followed by a survey of the light-cone current algebra of Fritzsch and Gell-Mann. The resulting derivations of the basic parton results, as well as the validity of the entire scheme in formal quark–gluon field theory, are examined. Moreover, the implications and consequences of the application of current algebra on light-planes are considered. The concluding chapter contains a discussion of the essential unity and dualism between the parton and light-cone approaches, with both critically using the idea of scale invariance. The omission of any discussion of other theoretical ideas on high-energy lepton–hadron reactions (such as vector dominance, resonance-summation, and Regge- or duality-based arguments) is deliberate.

This book grew out of lectures given at the Tata Institute of Fundamental Research during the summer of 1972. No systematic attempt has been made to give proper credit to workers who initially derived the results used here. The bibliography does, however, include most review articles and books written to date on this subject, where the original papers are quoted. Apologies are made to any reviewer whose name may have been inadvertently omitted. I am indebted to the graduate students of the Tata Institute of Fundamental Research for taking careful notes of my lectures, and especially to G. Bhattacharya, S. C. Chhajlany, and B. Radhakrishnan for their assistance in using them in the preparation of the manuscript. My understanding of the parton picture has been aided by conversations with S. Brodsky and T. M. Yan. My ideas on scale invariance have been shaped by inspiring discourses from K. Wilson, by insights gleaned from N. Mukunda, and by briefings on latest developments from G. Rajasekaran. My grasp of light-cone physics owes much to clarifications obtained from M. Gell-Mann, to comments made by T. Das, and to lectures given by L. K. Pandit. To all these I express my gratitude. Finally, it is a pleasure to thank V. Mubayi for his help in the publication of this book.

*Bombay, India* P.R.
*December 1973*

# CONTENTS

# NOTATION AND CONVENTIONS

WE use natural units $c = \hbar = 1$ and the metric, gamma matrices, and Dirac spinors as defined in *Relativistic quantum fields* by Bjorken and Drell (McGraw-Hill, 1965). The following additional information regarding our notation and conventions is given for the benefit of the reader.

Normalization of states: $\langle p'|p\rangle = \delta^{(3)}(\mathbf{p}'-\mathbf{p})$
Translation of operators: $O(x) = \mathrm{e}^{\mathrm{i}x.P}O(0)\,\mathrm{e}^{-\mathrm{i}x.P}$
Abbreviations:

| | | | |
|---|---|---|---|
| $l$ = | lepton, | $\gamma$ = | photon, |
| H = | hadron, | $\gamma^*$ = | virtual photon, |
| N = | nucleon, | $\pi$ = | pion, |
| e = | electron, | $\mu$ = | muon, |
| $p$ or p = | proton, | $\nu$ = | neutrino, |
| n = | neutron, | $\bar{\nu}$ = | antineutrino, |
| $\mathrm{q}_{\mathrm{p,n},\lambda}$ = | quarks, | $\bar{\mathrm{q}}_{\mathrm{p,n},\lambda}$ or $\mathrm{q}_{\bar{\mathrm{p}},\bar{\mathrm{n}},\bar{\lambda}}$ = | antiquarks. |

Standard deep inelastic parameters:

$q$ = leptonic (four-) momentum transfer,

$Q^2 = -q^2$, the negative squared mass carried by a current $J_\mu$,

$\nu = \dfrac{q\cdot p}{M}$, the leptonic energy transfer in the laboratory frame,

$W = \sqrt{\{(p+q)^2\}}$, the mass of the inelastic hadronic system,

$w$ = scale variable $\dfrac{Q^2}{2M\nu} = \omega^{-1}$,

$W_i(Q^2, \nu)$ = deep inelastic structure functions.

Throughout we denote a cross-section by the symbol $\sigma$ and a transition amplitude by $S$. SU(3) structure constants $f_{ijk}$ and $d_{ijk}$ and the matrices $\lambda^i$ are as defined in the book *The eightfold way* by Gell-Mann and Nee'man (Benjamin, 1964).

# INTRODUCTION

LEPTONS do not themselves have strong interactions. However, the quantum of weak or electromagnetic interaction generated by a lepton is able to interact with hadrons. It can either scatter from a hadronic target or produce hadrons directly, depending on the circumstances. The mechanisms involved make use of both the weak and electromagnetic properties of the hadrons and of the hadronic structure of the quantum itself. These complications are formally handled by means of hadronic currents which may be introduced through the following two interaction terms in the effective Lagrangian density:

$$\mathscr{L}_{\mathrm{I}}^{\mathrm{EM}} = e(j_{\mu}^{\mathrm{EM}} + J_{\mu}^{\mathrm{EM}})A^{\mu}, \qquad \mathscr{L}_{\mathrm{I}}^{\mathrm{W}} = \frac{G}{\sqrt{2}}(j_{\mu}^{\mathrm{W}} + J_{\mu}^{\mathrm{W}})(j^{\mu,\mathrm{W}} + J^{\mu,\mathrm{W}})^{\dagger}.$$

In the above equations $j_{\mu}^{\mathrm{EM,W}}$ and $J_{\mu}^{\mathrm{EM,W}}$ are the appropriate leptonic and hadronic currents respectively and $A_{\mu}$ is the electromagnetic field. There are two advantages in using lepton–hadron processes to probe the structure of hadrons. First, we may use perturbation theory for the part where the quantum of weak or electromagnetic interaction is generated by the lepton; for most purposes lowest-order considerations suffice. The second advantage is that, by appropriately varying the momentum transfer imparted by the lepton, we are able to vary the effective virtual mass associated with the current $J_{\mu}$; it follows that for large values (imaginary or real) of such masses we can obtain information on the short-distance structure of the hadrons.

The study of hadron structure by means of lepton-induced reactions was initiated with the simplest process, namely elastic scattering: $l + \mathrm{H} \rightarrow l' + \mathrm{H}'$. The two reactions of this type which have been of main interest are:

$$\mathrm{e} + \mathrm{N} \rightarrow \mathrm{e}' + \mathrm{N}' \text{ and } \begin{matrix} \nu_{\mu} \\ \bar{\nu}_{\mu} \end{matrix} + \mathrm{N} \rightarrow \mu^{\mp} + \mathrm{N}'.$$

The electromagnetic case has been studied experimentally (Hofstadter 1963) and theoretically (Drell and Zachariasen 1961) since the end of the 1950s. The weak reaction has attracted attention since the 1960s (Llewellyn-Smith 1972*c*). These investigations have sharpened our knowledge about certain properties of the nucleons, such as their magnetic and charge radii, apart from throwing light on the validity of the existing theories of weak and electromagnetic interactions. Another related process is the two-body production reaction $\mathrm{e}^{+} + \mathrm{e}^{-} \rightarrow \pi^{+} + \pi^{-}$ observed in storage rings (e.g. Martin 1967). The major problem with the study of all reactions such as these is that the differential cross-section becomes rather small when the magnitude of the squared mass $q^{2}$ of the corresponding hadronic current gets large. Hence, in this kinematic region of great interest, the extraction of a substantial amount of information from the data becomes difficult. On the other hand, inelastic lepton–nucleon scattering, namely, $l+\mathrm{N} \to l'+$ 'anything', does not show this behaviour. The same is true of the annihilation reaction $\mathrm{e}^+ + \mathrm{e}^- \to$ 'anything'. Here, 'anything' stands for any possible hadronic system produced. In the former reaction only the outgoing lepton is detected, and the sum is taken over all the hadrons. In the latter, those events which correspond only to hadronic final states are isolated and summed. In the limit when $q^2$ is very large in magnitude and 'anything' carries a very high mass, these reactions fall under the general category of deep inelastic lepton–hadron processes.

In this book we shall concentrate mainly on the reactions

$$\mathrm{e}+\mathrm{N} \to \mathrm{e}+\text{'anything' and } \begin{matrix}\nu_\mu \\ \bar{\nu}^\mu\end{matrix} + \mathrm{N} \to \mu^{\mp} + \text{'anything'}.$$

We shall also briefly consider the reaction $\mathrm{e}^+ + \mathrm{e}^- \to$ 'anything'. All these processes will be dealt with in the lowest order of perturbation theory for weak and electromagnetic interactions. These are the most basic reactions among deep inelastic lepton–hadron processes. There are many other processes which can also be given the same general heading, namely,

$$l+\mathrm{N} \to l'+\mathrm{H}+\text{'anything'},$$
$$\mathrm{e}^+ + \mathrm{e}^- \to \mathrm{H}+\text{'anything'},$$
$$\mathrm{H}+\mathrm{N} \to l^+ + l^- + \text{'anything'},$$

etc. However, the theoretical models used in attempts to understand these reactions are based on ideas acquired from the study of those mentioned earlier. Moreover, both the theoretical and experimental knowledge of these latter reactions are in a diffuse and fragmented state at present; therefore, we shall not consider them here. The interested reader may look up the review articles quoted in the bibliography for appropriate references.

## Preliminary kinematics and empirics

### *Elastic* ep *scattering*

We take the reaction $\mathrm{e}+\mathrm{p} \to \mathrm{e}'+\mathrm{p}'$ with single-photon exchange (Fig. 1). By four-momentum conservation, we have

$$p_\mathrm{e} + p = p'_\mathrm{e} + p'.$$

The scattering amplitude may be written (with $s$ referring to spin) as

$$S = \frac{1}{(2\pi)^3} \frac{m_\mathrm{e} e^2}{\sqrt{(p'_{\mathrm{e}0} p_{\mathrm{e}0})}} \frac{\bar{u}_{\mathrm{e}'} \gamma^\mu u_\mathrm{e}}{q^2 + \mathrm{i}\varepsilon} \langle p', s' | J_\mu^{\mathrm{EM}} | p, s \rangle \mathrm{i}(2\pi)^4 \delta^{(4)}(p' + p'_\mathrm{e} - p - p_\mathrm{e}),$$

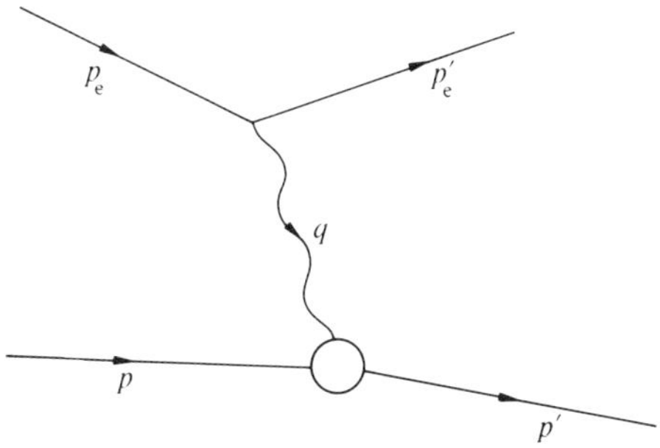


FIG. 1. Elastic electron–proton scattering by single-photon exchange.

where current conservation is maintained by requiring $\partial \,.\, J^{\mathrm{EM}} = 0$. The general form of the matrix element, consistent with parity, time-reversal invariance, current conservation and the hermiticity of the current $J_\mu^{\mathrm{EM}}$, is the following:

$$\langle p', s'|J_\mu^{\mathrm{EM}}|p, s\rangle = \frac{1}{(2\pi)^3}\frac{M}{\sqrt{(p_0 p'_0)}}\bar{u}_{s'}(p')\left\{F_1(q^2)\gamma_\mu + \mathrm{i}\frac{\kappa}{2M}F_2(q^2)\sigma_{\mu\nu}q^\nu\right\}u_s(p),$$

with the form factors $F_{1,2}(q^2)$ real and $F_{1,2}(0) = 1$. Here $\kappa$ is the anomalous magnetic moment of the proton. We have put $m_e$ for the mass of the electron and $M$ for the mass of the proton. From the above two equations we can calculate the differential cross-section and arrive at the Rosenbluth formula (in the laboratory frame),

$$\frac{\mathrm{d}\sigma}{\mathrm{d}\Omega_e} = \frac{\alpha^2}{4}\frac{[\{(F_1(q^2))^2 - (\kappa^2 q^2/4M^2)(F_2(q^2))^2\}\cos^2(\theta_e/2) - (q^2/2M^2)\{F_1(q^2) + \kappa F_2(q^2)\}^2\sin^2(\theta_e/2)]}{p_{e0}^2\sin^4(\theta_e/2)[1 + (2p_{e0}/M)\sin^2(\theta_e/2)]},$$

where $\mathrm{d}\Omega_e = 2\pi\,\mathrm{d}(\cos\theta_e)$, $e^2 = 4\pi\alpha$, $\theta_e$ is the laboratory scattering angle, and $(p'_e - p_e)^2 = q^2 \simeq -4p_{e0}p'_{e0}\sin^2(\theta_e/2)$. Here $q$ is a space-like vector, and we have neglected $m_e$ compared to $p_{e0}, p'_{e0}$ above. Let us remark here that, experimentally, it is the elastic form factors $F_{1,2}(q^2)$ that fall off very quickly for large increasing $|q^2|$.

*Inelastic* ep *scattering*

We now consider the reaction $\mathrm{e} + \mathrm{p} \to \mathrm{e}' +$ 'anything' (Fig. 2). The corresponding amplitude may be written (with the state $|n\rangle$ for 'anything') as

$$S = \left(\frac{1}{2\pi}\right)^3 \frac{m_e e^2}{\sqrt{(p_{e0}p'_{e0})}}\frac{\bar{u}_{e'}\gamma^\mu u_e}{q^2 + \mathrm{i}\varepsilon}\mathrm{i}(2\pi)^4\delta^{(4)}(p + q - p_n)\langle n|J_\mu^{\mathrm{EM}}|p, s\rangle. \qquad \text{(I.1)}$$

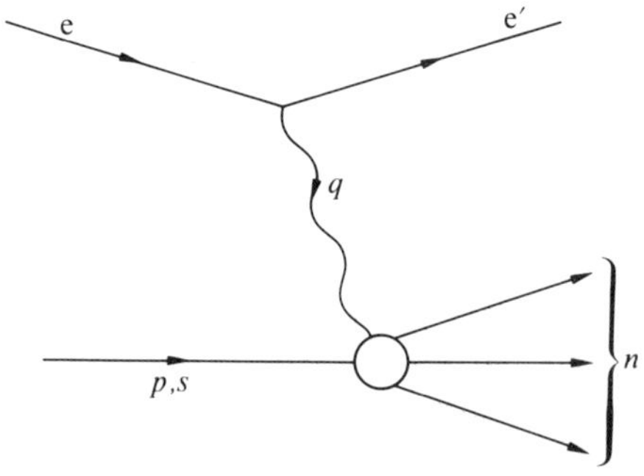


FIG. 2. Inelastic electron–proton scattering by single-photon exchange.

Ignoring terms containing the electron mass, we obtain

$$\frac{\mathrm{d}^2\sigma}{\mathrm{d}p'_{\mathrm{e}0}\,\mathrm{d}\Omega_{\mathrm{e}}} = \frac{\alpha^2}{q^4}\frac{p'_{\mathrm{e}0}}{p_{\mathrm{e}0}}\tfrac{1}{2}\mathrm{Tr}(\not{p}_{\mathrm{e}}\gamma^{\mu}\not{p}'_{\mathrm{e}}\gamma^{\nu})\sum_{n}\overline{\sum_{s}}\langle p,s|J^{\mathrm{EM}}_{\mu}|n\rangle\langle n|J^{\mathrm{EM}}_{\nu}|p,s\rangle\times$$
$$\times(2\pi)^6\delta^{(4)}(p+q-p_n). \qquad \text{(I.2)}$$

In eqn (I.2) the notation $\overline{\sum}_s$ implies that the spins of the initial proton are averaged. Henceforth we shall drop the spin index $s$ on the understanding that it is averaged out, and define $Q^2 \equiv -q^2$, $\nu \equiv (p\,.\,q/M)$ ($= q_0$ in the laboratory frame) and $W^2 \equiv p_n^2 = (p+q)^2 = M^2+2M\nu-Q^2 \rangle M^2$. Hence $w = Q^2/2M\nu$ is in the range $1 > w > 0$ and $\omega \equiv w^{-1}$ satisfies the inequality $1 < \omega < \infty$. Let us introduce the tensor

$$W^{\mathrm{e}}_{\mu\nu} \equiv \frac{p_0}{M}\sum_{n}\langle p|J^{\mathrm{EM}}_{\mu}|n\rangle\langle n|J^{\mathrm{EM}}_{\nu}|p\rangle(2\pi)^6\delta^{(4)}(p+q-p_n). \qquad \text{(I.3)}$$

P- and $T$-invariance imply

$$W^{\mathrm{e}}_{\mu\nu} = W^{\mathrm{e}}_{\nu\mu}. \qquad \text{(I.4)}$$

Then we have the general form

$$W^{\mathrm{e}}_{\mu\nu} = -g_{\mu\nu}W^{\mathrm{e}}_1+\frac{p_\mu p_\nu}{M^2}W^{\mathrm{e}}_2+\frac{1}{M^2}(p_\mu q_\nu+p_\nu q_\mu)W^{\mathrm{e}}_4+\frac{q_\mu q_\nu}{M^2}W^{\mathrm{e}}_5, \qquad \text{(I.5)}$$

where $W^{\mathrm{e}}_i \equiv W^{\mathrm{e}}_i(Q^2,\nu)$ are real. Current conservation requires $q^\mu W^{\mathrm{e}}_{\mu\nu} = 0 = q^\nu W^{\mathrm{e}}_{\mu\nu}$. This, in turn gives the following relations:

$$W^{\mathrm{e}}_4 = -\frac{p\,.\,q}{q^2}W^{\mathrm{e}}_2 \quad \text{and} \quad W^{\mathrm{e}}_5 = \frac{M^2}{q^2}W^{\mathrm{e}}_1+\frac{(p\,.\,q)^2}{q^4}W^{\mathrm{e}}_2. \qquad \text{(I.6)}$$

From eqns (I.5) and (I.6) is obtained the famous formula

$$W^{\mathrm{e}}_{\mu\nu} = \left(-g_{\mu\nu}+\frac{q_\mu q_\nu}{q^2}\right)W^{\mathrm{e}}_1+\frac{1}{M^2}\left(p_\mu-\frac{p\,.\,q}{q^2}\right)\left(p_\nu-\frac{p\,.\,q}{q^2}q_\nu\right)W^{\mathrm{e}}_2, \qquad \text{(I.7)}$$

where $W^{\mathrm{e}}_{1,2}$ are called inelastic ep structure functions.

Eqn (I.3) may be written as

$$W^{\mathrm{e}}_{\mu\nu} = \frac{1}{2\pi}\int \mathrm{d}^4x\, \mathrm{e}^{\mathrm{i}q.x}\langle p|J^{\mathrm{EM}}_{\mu}(x)J^{\mathrm{EM}}_{\nu}(0)|p\rangle \frac{p_0}{M}(2\pi)^3. \tag{I.8}$$

In contrast, we have

$$\frac{1}{2\pi}\int \mathrm{d}^4x\, \mathrm{e}^{\mathrm{i}q.x}\langle p|J^{\mathrm{EM}}_{\nu}(0)J^{\mathrm{EM}}_{\mu}(x)|p\rangle \frac{p_0}{M}(2\pi)^3 = 0. \tag{I.9}$$

Eqn (I.9) follows because, on using the translation property, the left-hand side can be shown to be proportional to $\delta^{(4)}(p_n+q-p)$; this vanishes since, with $q_0 > 0$ and $p_{n0} > p_0$, $(p_n+q)_0 > p_0$. Thus from eqns (I.8) and (I.9) it follows that

$$W^{\mathrm{e}}_{\mu\nu} = \frac{1}{2\pi}\frac{p_0}{M}(2\pi)^3 \int \mathrm{d}^4x\, \mathrm{e}^{\mathrm{i}q.x}\langle p|[J^{\mathrm{EM}}_{\mu}(x), J^{\mathrm{EM}}_{\nu}(0)]|p\rangle. \tag{I.10}$$

Eqn (I.10) is interesting from a theoretical point of view in that the tensor of interest has been related to a commutator of two electromagnetic currents. From this equation it is straightforward to obtain the 'crossing' relation

$$W^{\mathrm{e}}_{\mu\nu}(-q, p) = -W^{\mathrm{e}}_{\mu\nu}(q, p). \tag{I.11}$$

Eqn (I.11) implies that $W^{\mathrm{e}}_{1,2}$ are odd functions of $\nu$, namely,

$$W^{\mathrm{e}}_{1,2}(q^2, \nu) = -W^{\mathrm{e}}_{1,2}(q^2, -\nu). \tag{I.12}$$

We can now calculate the differential cross-section in the laboratory frame. With

$$\mathrm{Tr}(\not p_{\mathrm{e}}\gamma^{\mu}\not p'_{\mathrm{e}}\gamma^{\nu}) = 4\left(p^{\mu}_{\mathrm{e}}p'^{\nu}_{\mathrm{e}} + p'^{\mu}_{\mathrm{e}}p^{\nu}_{\mathrm{e}} + \frac{q^2}{2}g^{\mu\nu}\right)$$

and eqn (I.7), we have, from eqn (I.2), the result

$$\begin{aligned}\frac{\mathrm{d}^2\sigma}{\mathrm{d}p'_{\mathrm{e}0}\,\mathrm{d}\Omega_{\mathrm{e}}} &= \frac{2\alpha^2}{Q^4}\frac{p'_{\mathrm{e}0}}{p_{\mathrm{e}0}}\left\{2W^{\mathrm{e}}_1 p_{\mathrm{e}}\cdot p'_{\mathrm{e}} + \frac{W^{\mathrm{e}}_2}{M^2}(2p\cdot p_{\mathrm{e}}p\cdot p'_{\mathrm{e}} - M^2 p_{\mathrm{e}}\cdot p'_{\mathrm{e}})\right\}\\ &= \frac{\alpha^2}{4p^2_{\mathrm{e}0}\sin^4(\theta_{\mathrm{e}}/2)}\left\{2W^{\mathrm{e}}_1(q^2,\nu)\sin^2\frac{\theta_{\mathrm{e}}}{2} + W^{\mathrm{e}}_2(q^2,\nu)\cos^2\frac{\theta_{\mathrm{e}}}{2}\right\}.\end{aligned} \tag{I.13}$$

In eqn (I.13) we have used $\mathrm{d}\Omega_{\mathrm{e}} = 2\pi\,\mathrm{d}(\cos\theta_{\mathrm{e}})$, $Q^2 \simeq 4p_{\mathrm{e}0}p'_{\mathrm{e}0}\sin^2(\theta_{\mathrm{e}}/2)$, and $\nu = p_{\mathrm{e}0} - p'_{\mathrm{e}0}$. This equation can be rewritten as

$$\frac{\mathrm{d}^2\sigma}{\mathrm{d}Q^2\,\mathrm{d}\nu} = \frac{\pi}{p_{\mathrm{e}0}p'_{\mathrm{e}0}}\frac{\mathrm{d}^2\sigma}{\mathrm{d}p_{\mathrm{e}0}\,\mathrm{d}p'_{\mathrm{e}0}} = \frac{4\pi\alpha^2}{Q^4}\frac{p'_{\mathrm{e}0}}{p_{\mathrm{e}0}}\left(2W^{\mathrm{e}}_1\sin^2\frac{\theta_{\mathrm{e}}}{2} + W^{\mathrm{e}}_2\cos^2\frac{\theta_{\mathrm{e}}}{2}\right). \tag{I.14}$$

We can re-interpret $W^{\mathrm{e}}_1$ and $W^{\mathrm{e}}_2$ in terms of appropriate virtual photo-absorption total cross-sections. If $\gamma^*$ stands for a virtual photon, consider

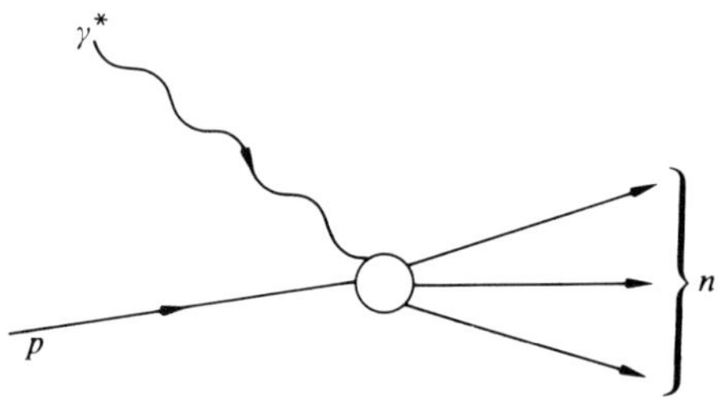


FIG. 3. Virtual photon absorption by a proton.

the reaction $\gamma^* + \mathrm{p} \to$ 'anything' (Fig. 3) in the laboratory frame, where $p = (m, 0, 0, 0)$ and $q = (\nu, 0, 0, \sqrt{(\nu^2 + Q^2)})$. We can visualize a beam of space-like photons of velocity $|\mathbf{q}|/q_0 = \sqrt{(\nu^2 + Q^2)}/\nu$ being absorbed by a proton going into a final state $|n\rangle$. The cross-section for the absorption of virtual photons on protons may then be written as

$$\sigma_{\gamma^*\mathrm{p}} = \frac{\alpha}{\sqrt{(\nu^2+Q^2)}} \sum_n (2\pi)^8 \varepsilon^{*\mu}(q) \varepsilon^{\nu}(q) \langle p|J_\mu^{\mathrm{EM}}|n\rangle \langle n|J_\nu^{\mathrm{EM}}|p\rangle \delta^{(4)}(p+q-p_n), \qquad (\mathrm{I.15})$$

where $\varepsilon$ refers to the polarization of the virtual photon. We can use the relation $\sum_\varepsilon \varepsilon^\mu \varepsilon^\nu = -g^{\mu\nu} + (q^\mu q^\nu / q^2)$ and current conservation $q \,.\, J = 0$ in the laboratory frame (i.e. $J_\parallel^{\mathrm{EM}} = \nu/\sqrt{(\nu^2+Q^2)} J_0^{\mathrm{EM}}$, where the subscripts $\parallel$ and $\perp$ refer respectively to components parallel and perpendicular to $\mathbf{q}$). This leads to

$$\sigma_{\gamma^*\mathrm{p}} = \sigma_\mathrm{T} + \sigma_\mathrm{L},$$

where the transverse and longitudinal virtual photo-absorption cross-sections $\sigma_\mathrm{T}$ and $\sigma_\mathrm{L}$ may be expressed as follows:

$$\sigma_\mathrm{T} = \frac{\alpha}{\sqrt{(\nu^2+Q^2)}} (2\pi)^8 \sum_n |\langle n|J_\perp^{\mathrm{EM}}|p\rangle|^2 \delta^{(4)}(p+q-p_n), \qquad (\mathrm{I.16})$$

$$\sigma_\mathrm{L} = \frac{\alpha}{\sqrt{(\nu^2+Q^2)}} \frac{Q^2}{\nu^2} (2\pi)^8 \sum_n |\langle n|J_\parallel^{\mathrm{EM}}|p\rangle|^2 \delta^{(4)}(p+q-p_n). \qquad (\mathrm{I.17})$$

From eqns (I.3), (I.7), (I.16), and (I.17) we can evaluate $\sigma_{\mathrm{T,L}}$ and relate them to $W_{1,2}^\mathrm{e}$. We thus obtain

$$W_1^\mathrm{e} = \frac{\sqrt{(\nu^2+Q^2)}}{4\pi^2\alpha} \sigma_\mathrm{T}, \qquad (\mathrm{I.18})$$

$$W_2^\mathrm{e} = \frac{\sqrt{(\nu^2+Q^2)}}{4\pi^2\alpha} \frac{Q^2}{(\nu^2+Q^2)} (\sigma_\mathrm{T} + \sigma_\mathrm{L}), \qquad (\mathrm{I.19})$$

which show from the positivity of the cross-sections that

$$W_1^\mathrm{e} \geqslant 0 \quad \text{and} \quad W_2^\mathrm{e}\left(1 + \frac{\nu^2}{Q^2}\right) \geqslant W_1^\mathrm{e}. \qquad (\mathrm{I.20})$$

*Inelastic* $(\nu, \bar{\nu})$p *scattering*

The relevant reactions are

$$\nu_\mu + \mathrm{p} \to \mu^- + \text{`anything'}, \qquad \nu_e + \mathrm{p} \to e^- + \text{`anything'},$$
$$\bar{\nu}_\mu + \mathrm{p} \to \mu^+ + \text{`anything'}, \qquad \bar{\nu}_e + \mathrm{p} \to e^+ + \text{`anything'}.$$

Experimentally, muon-neutrino beams are more readily available, so that henceforth we shall use the notation $\nu$ to denote the muon neutrino. First take the $\nu$p case (Fig. 4). With $q = p_\nu - p_\mu$, the scattering amplitude has the form

$$S = \lim_{m_\nu \to 0} \left(\frac{1}{2\pi}\right)^3 \sqrt{\left(\frac{m_\mu m_\nu}{p_{\mu 0} p_{\nu 0}}\right)} \frac{G}{\sqrt{2}} \times$$
$$\times \bar{u}_\mu \gamma^\alpha (1-\gamma_5) u_\nu \mathrm{i}(2\pi)^4 \delta^{(4)}(p+q-p_n) \langle n | J_\alpha^{\mathrm{W}} | p, s \rangle. \tag{I.21}$$

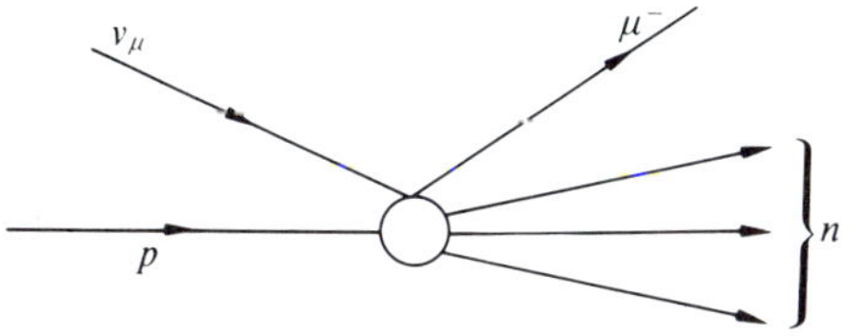


FIG 4. Inelastic neutrino–proton interaction.

As before, the differential cross-section may be written as

$$\frac{\mathrm{d}^2\sigma^{(\nu)}}{\mathrm{d}p'_{\mu 0}\,\mathrm{d}\Omega_\mu} = \left(\frac{G}{4\pi\sqrt{2}}\right)^2 \frac{p_{\mu 0}}{p_{\nu 0}} \operatorname{Tr}\{\not{p}_\nu \gamma^\alpha (1-\gamma_5) \not{p}_\mu \gamma^\beta (1-\gamma_5)\} W^\nu_{\alpha\beta} \frac{M}{p_0}$$
$$\simeq \frac{G^2}{4\pi^2} \frac{p_{\mu 0}}{p_{\nu 0}} (p_\nu^\alpha p_\mu^\beta + p_\nu^\beta p_\mu^\alpha - g^{\alpha\beta} p_\nu \cdot p_\mu - \mathrm{i}\varepsilon^{\alpha\beta\gamma\delta} p_{\nu\gamma} p_{\mu\delta}) W^\nu_{\alpha\beta} \frac{M}{p_0} \tag{I.22}$$

in the approximation $m_\mu / p_{\mu 0} \to 0$. In eqn (I.22), we have introduced a structure tensor in analogy with eqn (I.10). Thus

$$W^\nu_{\alpha\beta} = \frac{1}{2\pi} \int \mathrm{d}^4x\, \mathrm{e}^{\mathrm{i}q.x} \langle p | [J_\alpha^{\mathrm{W}\dagger}(x), J_\beta^{\mathrm{W}}(0)] | p \rangle (2\pi)^3 \frac{p_0}{M}. \tag{I.23}$$

Similarly in the $(\bar{\nu}_\mu, \mu^+)$ case, we obtain the expression

$$\frac{\mathrm{d}^2\sigma^{(\bar{\nu})}}{\mathrm{d}p'_{\bar{\mu}0}\,\mathrm{d}\Omega_{\bar{\mu}}} = \frac{G^2}{4\pi^2} \frac{p_{\bar{\mu}0}}{p_{\bar{\nu}0}} (p_{\bar{\nu}}^\alpha p_{\bar{\mu}}^\beta + p_{\bar{\nu}}^\beta p_{\bar{\mu}}^\alpha - g^{\alpha\beta} p_{\bar{\nu}} \cdot p_{\bar{\mu}} + \mathrm{i}\varepsilon^{\alpha\beta\gamma\delta} p_{\bar{\nu}\gamma} p_{\bar{\mu}\delta}) \frac{M}{p_0} W^{\bar{\nu}}_{\alpha\beta}, \tag{I.24}$$

where

$$W^{\bar{\nu}}_{\alpha\beta} = \frac{1}{2\pi} \int \mathrm{d}^4x\, \mathrm{e}^{\mathrm{i}q.x} \langle p | [J_\alpha^{\mathrm{W}}(x), J_\beta^{\mathrm{W}\dagger}(0)] | p \rangle (2\pi)^3 \frac{p_0}{M}. \tag{I.25}$$

In analogy with eqn (I.11), we now have

$$W^{\bar{\nu}}_{\beta\alpha}(-q, p) = -W^{\nu}_{\alpha\beta}(q, p). \tag{I.26}$$

The general form of these tensors is a little more complicated than in the ep case because of parity violation and the non-hermiticity of the weak currents. Here we can write

$$W^{\nu\bar{\nu}}_{\alpha\beta} = -g_{\alpha\beta}W^{\nu\bar{\nu}}_1 + \frac{p_\alpha p_\beta}{M^2}W^{\nu\bar{\nu}}_2 - \frac{\mathrm{i}}{2M^2}\varepsilon_{\alpha\beta\gamma\delta}p^\gamma q^\delta W^{\nu\bar{\nu}}_3 +$$

$$+\frac{1}{2M^2}(p_\alpha q_\beta + p_\beta q_\alpha)W^{\nu\bar{\nu}}_4 + \frac{1}{M^2}q_\alpha q_\beta W^{\nu\bar{\nu}}_5 + \frac{\mathrm{i}}{2M^2}(p_\alpha q_\beta - p_\beta q_\alpha)W^{\nu\bar{\nu}}_6, \tag{I.27}$$

where $W^{\nu\bar{\nu}}_i \equiv W^{\nu\bar{\nu}}_i(q^2, \nu)$. The third term in the right-hand side of eqn (I.27) demonstrates that parity is violated in the process.

It is clear that, when eqn (I.27) is substituted in eqn (I.22) or eqn (I.24), $W_6$ does not contribute and the functions $W_4$ and $W_5$ pick up terms proportional to $m_\mu$, which we neglect. Thus we need consider only $W_{1,2,3}$, all of which may be seen to be real because of the relation $W^*_{\alpha\beta} = W_{\beta\alpha}$. Finally then, we obtain in the laboratory frame the formulae

$$\frac{\mathrm{d}^2\sigma^{(\nu)}}{\mathrm{d}Q^2\,\mathrm{d}\nu} = \frac{G^2}{2\pi}\frac{p_{\mu 0}}{p_{\nu 0}}\left(2W^{\nu}_1 \sin^2\frac{\theta_\mu}{2} + W^{\nu}_2\cos^2\frac{\theta_\mu}{2} + W^{\nu}_3\frac{p_{\mu 0}+p_{\nu 0}}{M}\sin^2\frac{\theta_\mu}{2}\right), \tag{I.28}$$

$$\frac{\mathrm{d}^2\sigma^{(\bar{\nu})}}{\mathrm{d}Q^2\,\mathrm{d}\nu} = \frac{G^2}{2\pi}\frac{p_{\bar{\mu} 0}}{p_{\bar{\nu} 0}}\left(2W^{\bar{\nu}}_1 \sin^2\frac{\theta_{\bar{\mu}}}{2} + W^{\bar{\nu}}_2\cos^2\frac{\theta_{\bar{\mu}}}{2} - W^{\bar{\nu}}_3\frac{p_{\bar{\mu} 0}+p_{\bar{\nu} 0}}{M}\sin^2\frac{\theta_{\bar{\mu}}}{2}\right). \tag{I.29}$$

It should be noted that, from eqns (I.26) and (I.27), we obtain

$$W^{\nu}_{1,2}(q^2, -\nu) \pm W^{\bar{\nu}}_{1,2}(q^2, -\nu) = \mp\{W^{\nu}_{1,2}(q^2, \nu) \pm W^{\bar{\nu}}_{1,2}(q^2, \nu)\} \tag{I.30}$$

and

$$W^{\nu}_3(q^2, -\nu) \pm W^{\bar{\nu}}_3(q^2, -\nu) = \pm\{W^{\nu}_3(q^2, \nu) \pm W^{\bar{\nu}}_3(q^2, \nu)\}. \tag{I.31}$$

Eqns (I.30) and (I.31) demonstrate the appropriate combinations of neutrino and antineutrino structure functions which are odd or even when $q$ changes sign. Positivity inequalities, analogous to eqn (I.20), can be derived also on these structure functions (see Exercise I.1).

*Deep inelastic behaviour and scaling*

We present the facts relevant to this part through the following comments.

1. Recall that the deep inelastic region in the reaction $l + \mathrm{N} \rightarrow l' +$ 'anything' refers to the domain where both $Q^2$ and $W^2 \equiv M^2 + 2M\nu - Q^2$ are large compared to $M^2_{\mathrm{H}} \simeq 1\ \mathrm{GeV}^2$. In lepton-induced resonance production $W$ becomes $M_{\mathrm{R}}$—the mass of the resonance (e.g. for N*-production $W = M_{\mathrm{N}*}$). Figure 5 illustrates the electro-production of a resonance with

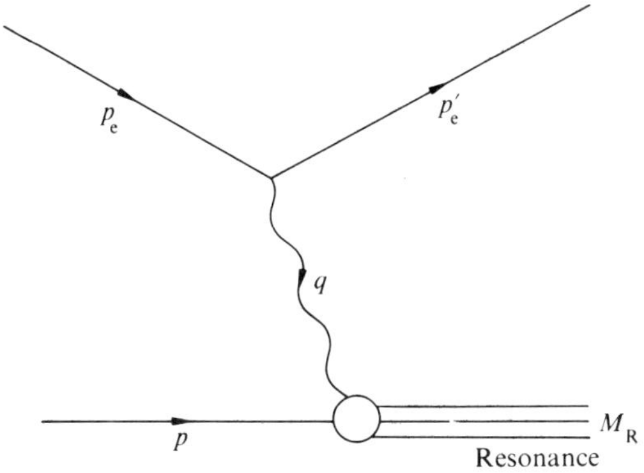


FIG. 5. Electro-resonance production by single-photon exchange.

single photon exchange. The curve in Fig. 6 is a typical (Taylor 1969) experimental curve for the differential cross-section with $Q^2 \gtrsim 1\ \text{GeV}^2$ and shows bumps in the resonance region. It is then smoothed out.

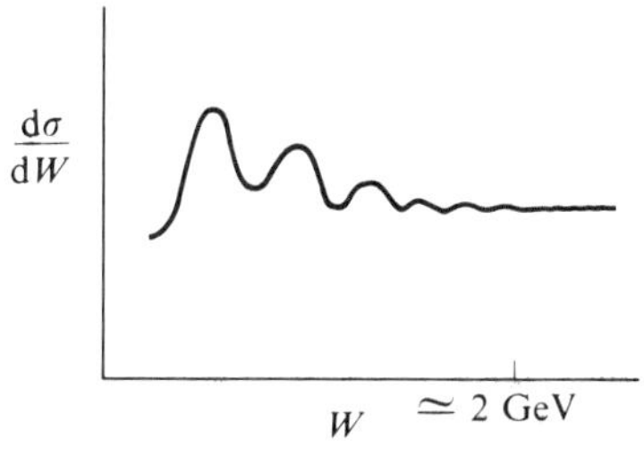


FIG. 6. Schematic representation of the transition from resonance production to the deep inelastic régime in lepton-induced reactions.

This 'smoothing' marks the onset of the deep inelastic domain. A two-dimensional plot with $Q^2$ and $2M\nu$ can be made, as shown in Fig. 7. In the case of a single resonance being produced, $M_R^2 = M^2 + 2M\nu - Q^2$, and for

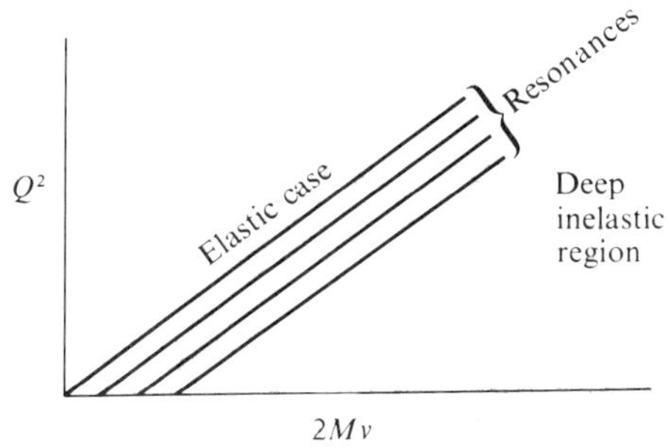


FIG. 7. $Q^2$ versus $2M\nu$ behaviour in inelastic lepton scattering.

the elastic case, $Q^2 = 2M\nu$. For a fixed and large value of $Q^2$, we move into the deep inelastic region as $2M\nu$ increases.

2. There is a striking difference in the behaviour of the elastic form factors and the deep inelastic structure functions at large $Q^2$. Take ep scattering, for instance. As mentioned earlier, the elastic form factors fall off very fast as $Q^2$ increases. Thus Fig. 8(a) shows that $F_1^{\text{ep}}$ experimentally tends to zero at least as fast as $1/Q^4$ for large $Q^2$. This is regarded as an effect of the structure of the proton due to strong interactions; if the electron saw the proton as a point particle, $F_1$ should have remained at unity for all $Q^2$. On the other hand, for the inelastic case the situation is shown (Miller *et al.* 1972) in Fig. 8(b) for $\nu \simeq 20$ GeV. This shows that the inelastic structure function $\nu W_2$ is fairly large for large $Q^2$ and is nearly constant there. It is exhibiting, so to speak, a point-like structure in the behaviour of the proton in deep inelastic scattering. It should be noted that the vanishing at $Q^2 = 0$ is kinematical in origin, as is clear from eqn (I.19). Similar large cross-sections are indicated from the preliminary data on $e^+ + e^- \to$ 'anything' (Silvestrini 1972).

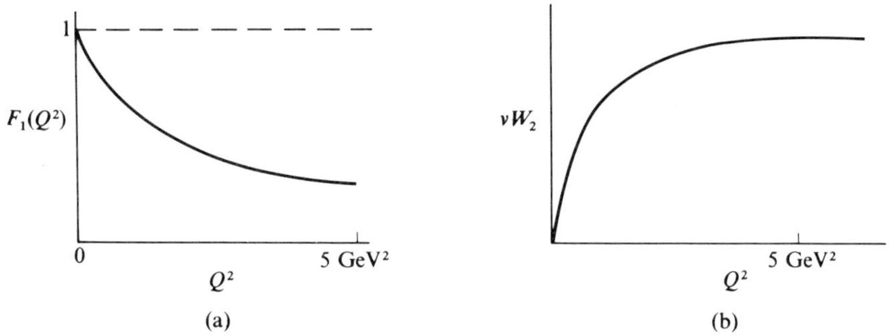


FIG. 8. $F_1(Q^2)$ (a) and $\nu W_2$ (b) versus $Q^2$ behaviour in electron–proton scattering.

3. The data on $W_1^{\text{ep}}$ and $\nu W_2^{\text{ep}}$ in the deep inelastic domain also demonstrate the scaling phenomenon, as shown in Fig. 9 (Taylor 1969; Kendall 1971). Scaling relations were originally proposed on theoretical grounds (Bjorken 1969) for the inelastic lepton–nucleon structure functions. According to these, in the limit (hereafter called the Bjorken or Bj limit) when $Q^2$ and $\nu \to \infty$ with $w \equiv Q^2/2M\nu$ (or equivalently $\omega \equiv w^{-1}$) fixed, the structure functions behave in the following way:

$$\lim_{\text{Bj}} W_1^{\text{e}\nu\bar{\nu}}(Q^2, \nu) = M^{-1}F_1(\omega), \tag{I.32a}$$

$$\lim_{\text{Bj}} \nu W_2^{\text{e}\nu\bar{\nu}}(Q^2, \nu) = F_2(\omega), \tag{I.32b}$$

$$\lim_{\text{Bj}} \nu W_3^{\nu\bar{\nu}}(Q^2, \nu) = F_3(\omega). \tag{I.32c}$$

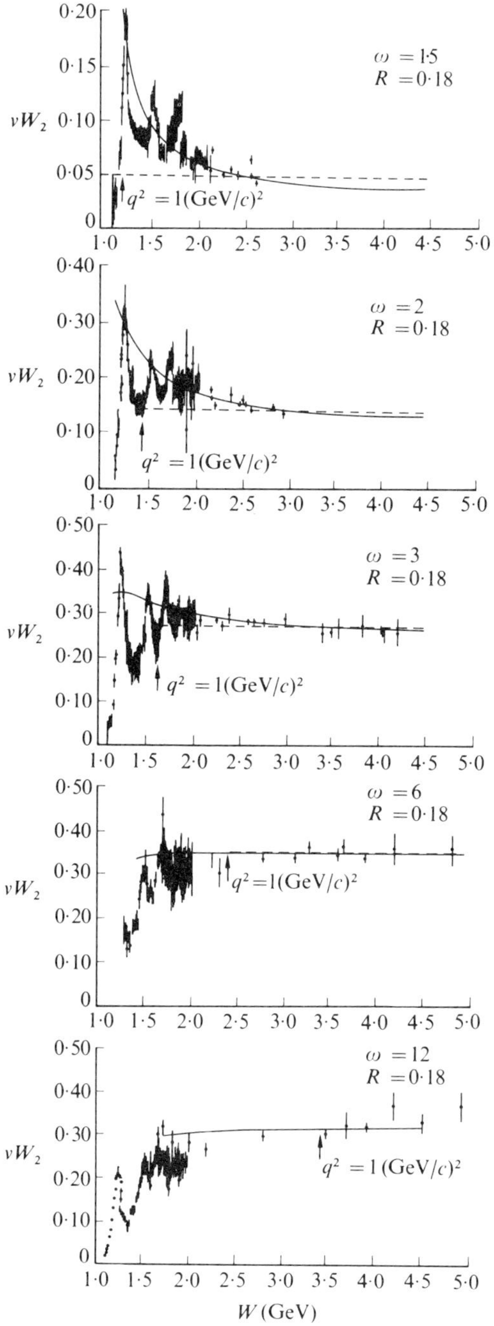


FIG. 9. Experimental evidence of Bjorken scaling in electron–proton scattering (Miller *et al.* 1972).

In eqns (I.32) $F_{1,2,3}(\omega)$ are functions only of the scale variable $\omega$. The experimental data obtained on inelastic eN scattering so far agree remarkably well with Bjorken's scaling predictions for $\nu \gtrsim 2{\cdot}5$ GeV and $Q^2 \gtrsim 1$ GeV$^2$. The predictions for the weak case are yet to be tested directly. It should be added that, despite immense difficulties, it is possible in principle to measure the structure functions $W_{4,5}^{\nu\bar{\nu}}$ in inelastic neutrino scattering. Scaling relations have been proposed for these too (e.g. Brown (1972) and Fayyazuddin and Riazuddin (1972)). However, they seem to depend sensitively on the details of the breaking of dilation invariance of the Hamiltonian density for strong interactions. The relations are of the form

$$\lim_{\mathrm{Bj}} \nu^{\alpha} W_{4,5}^{\nu\bar{\nu}}(Q^2, \nu) = F_{4,5}(\omega),$$

where, by general agreement, $1 \leqslant \alpha \leqslant 3$. We shall not consider the structure functions $W_{4,5}$ any further in this book, except to mention here that symmetry relations among them have been obtained (Llewellyn-Smith 1972c).

## Exercises

I.1. In analogy with eqn (I.20), derive the positivity inequality

$$\frac{1}{2M}\sqrt{(\nu^2+Q^2)}|W_3^{\nu\bar{\nu}}| \leqslant W_1^{\nu\bar{\nu}} \leqslant W_2^{\nu\bar{\nu}}\left(1+\frac{\nu^2}{Q^2}\right).$$

I.2. Starting from eqn (I.27) and keeping muon mass terms, derive more general expressions for the differential cross-sections of eqns (I.28) and (I.29).

PART I

PARTONS

# 1

# THE KINDERGARTEN PARTON MODEL

## Basic ideas

THE parton model (Feynman 1969, 1972; Bjorken and Paschos 1969; Drell 1969) is a set of approximations which are supposed to apply in deep inelastic lepton–hadron processes. Parton ideas have been used also in the context of hadron–hadron collisions (Feynman 1969; Kogut and Susskind 1973; Gunion, Brodsky, and Blankenbecler 1972, 1973); however, we are not concerned with that approach here. The motivation for the parton concept came from the need to understand the marked difference in the behaviour of the elastic and the inelastic differential cross-sections for electron–proton scattering at large momentum transfers. In the elastic ep case we saw something completely different from a point-like behaviour, whereas the deep inelastic case showed the presence of point-like interactions. Note that the elastic form factors were defined through the matrix element $\langle p'|J_\mu^{\mathrm{EM}}|p\rangle$, while the inelastic structure functions were introduced in terms of sums of expressions such as $\sum_n |\langle n|J^{\mathrm{EM}}|p\rangle|^2$, or in other words in terms of $\sigma_{\mathrm{T,L}}^{\gamma^*\mathrm{p}}$. A diagrammatic representation for the former is given in Fig. 10, while that for the latter appears in Fig. 11.

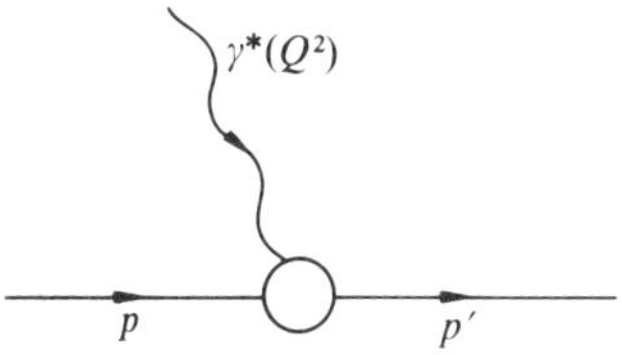


FIG. 10. Elastic form factor for electron–proton scattering.

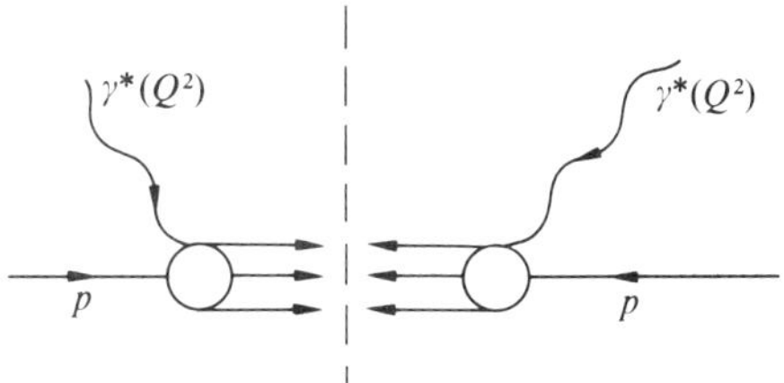


FIG. 11. Inelastic structure function for electron–proton scattering.

Qualitatively, the fact that the behaviour of $\nu W_2$ indicates point-like interaction suggests that the situation in deep inelastic scattering is as shown in Fig. 12. We guess that, since in this case we are adding probabilities, the total probability becomes large in the asymptotic region; the probability

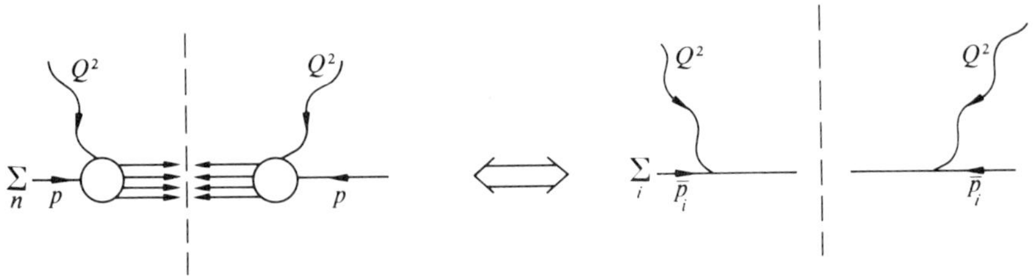


FIG. 12. Equivalence of sum over intermediate states and incoherent summation of scattering from individual point-like constituents of the proton in deep inelastic electron–proton scattering.

for the case of Fig. 10 is small, because here amplitudes have to be added. Figure 12 suggests the replacement of the sum over intermediate states $|n\rangle$ by the incoherent scattering of $\gamma^*$ from individual point-like constituents of the proton. In contrast, we can argue that coherence effects overwhelm the scattering of $\gamma^*$ from point-like constituents in Fig. 10.

The serious and quantitative construction of any constituent model, based on the idea of the impulse approximation, should be preceded by an examination of inelastic non-relativistic electron scattering in the Coulomb field of the nucleus in search of clues. The amplitude for the last-mentioned process (Fig. 13) is proportional to

$$\frac{e^2}{|\mathbf{q}|^2}\left\langle X \,\middle|\, \sum_{\text{protons}} \exp(\mathrm{i}\mathbf{q}\,.\,\mathbf{r}_i) \,\middle|\, \text{nucleus} \right\rangle,$$

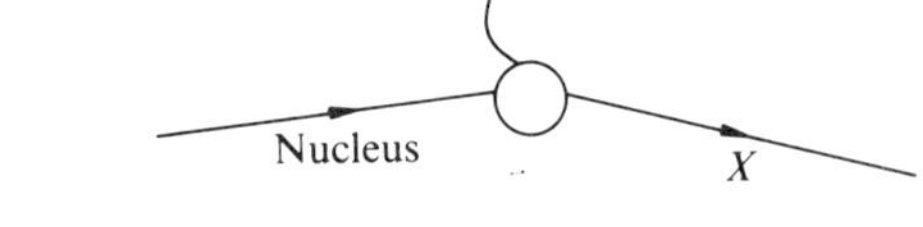

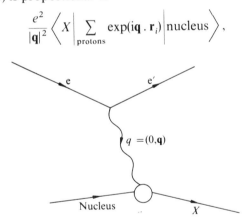

FIG. 13. Inelastic electron–nucleus scattering.

where $\mathbf{r}_i$ denotes the position of the ith proton. Hence

$$\int \mathrm{d}\nu \frac{\mathrm{d}^2\sigma}{\mathrm{d}|\mathbf{q}|^2\mathrm{d}\nu} = \frac{4\pi\alpha^2}{|\mathbf{q}|^4}\left\langle \text{nucleus} \,\middle|\, \sum_{i,j} \exp\{\mathrm{i}\mathbf{q}\,.\,(\mathbf{r}_i - \mathbf{r}_j)\} \,\middle|\, \text{nucleus} \right\rangle$$

$$= \frac{4\pi\alpha^2}{|\mathbf{q}|^4}\{Z + Z(Z-1)f(|\mathbf{q}|^2)\},$$

where $Z$ is the atomic number. The first term within the square bracket of the above equation comes from the incoherent part, while the term $Z(Z-1)f(|\mathbf{q}|^2)$ comes from the coherent part with $f(|\mathbf{q}|^2)$ as the correlation function. Experimentally, $f(|\mathbf{q}|^2)$ vanishes if $|\mathbf{q}| \gg$ (mean internucleon separation)$^{-1} \simeq 150$ MeV. This means that the incoherent part makes a good approximation to the cross-section when $Q^2$ is large, i.e. $Q^2 \gg (\Delta r_{\mathrm{N}})^{-2}$, where $\Delta r_{\mathrm{N}}$ is the internucleon separation. Thus we know from this example that the impulse approximation works when the momentum transfer is sufficiently large to allow us to neglect the coherent part and retain only the incoherent part. We should like to try a similar approach in inelastic electron–proton scattering, where the proton will be treated as a gas of free 'partons'—the basic point-like constituents of the proton.

Let us see why the impulse approximation works in the atomic and nuclear cases even for not too high energies. If we study ratios of the binding energy involved to the rest energy of the constituent in an atom and in a nucleus, we see that

$$\frac{\text{binding energy}}{\text{rest energy}} \sim \frac{\text{few eV}}{0{\cdot}51\ \text{MeV}} \sim 10^{-5} \ll 1 \quad \text{for an atomic electron}$$

$$\sim \frac{10\ \text{MeV}}{940\ \text{MeV}} \sim 10^{-2} \ll 1 \quad \text{for a nuclear proton or neutron.}$$

The same ratio in the case of partons in a proton is not expected to be much less than unity, creating an immediate difficulty. Further, if the impulse approximation has to work, the time of interaction (between the virtual photon and the parton) $\tau$ has to be much less than the lifetime $T$ of the virtual constituents (i.e. partons). Now we note that both the binding energy and the lifetime $T$ are frame-dependent notions. Feynman suggested that we should go to the infinite momentum frame (i.e. where the $z$-component of momentum $p_z \to \infty$). There, because of time dilation, the virtual constituents of the proton are longer lived and freer. Hence if the interaction time $\tau$ is much less than $T$ in this frame, with $Q^2$ being large, the scattering could be taken to be instantaneous as well as incoherent.

In an infinite momentum frame the proton deforms into a pancake, suffering longitudinal contraction, whereas the transverse dimensions are not affected. Furthermore, the relative velocities among the partons are slowed down. If now the virtual photon propagates mainly in the transverse direction with a large transverse momentum, then the scattered parton will speed away effectively free from the interaction of the rest in a relatively transverse direction. As a crude estimate of the condition under which the parton model should work, consider the following old-fashioned perturbation calculation. Take the virtual states on the mass shell and energy as not quite conserved (although momentum is) in the transition to the intermediate

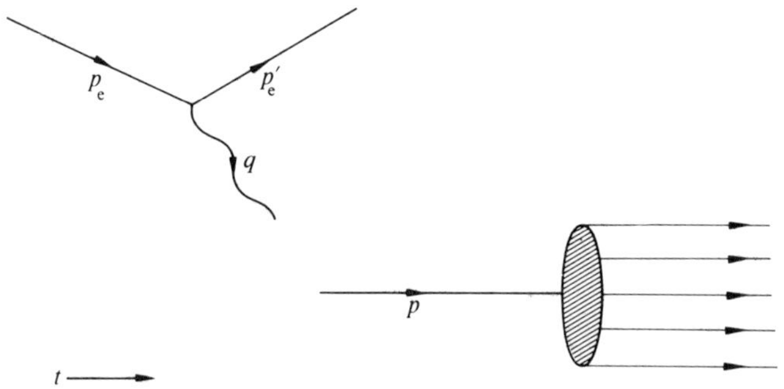


FIG. 14. Simple perturbation theory representation of the parton model.

states (Fig. 14). The interaction time of the virtual photon is $\tau \simeq (p'_{e0} - p_{e0})^{-1} = q_0^{-1}$. The lifetime of the virtual states (of which $i$ is a particular constituent) is ($E_i$ and $E_p$ being the energies of the $i$th parton and of the proton respectively)

$$T \simeq \frac{1}{\Delta E} = \frac{1}{\sum_i E_i - E_p}$$

$$= \frac{1}{\sum_i \{(x_i \mathbf{P})^2 + p_{i\perp}^2 + \mu_i^2\}^{\frac{1}{2}} - \sqrt{(P^2+M^2)}}. \tag{1.1}$$

In eqn (1.1) $p_{iz} = x_i P$, $\sum_i x_i = 1$, and $\sum_i \mathbf{p}_{i\perp} = 0$; here $\mathbf{p}_i$ refers to the momentum of the $i$th parton and $x_i$ is its fraction of the total longitudinal momentum. Let us choose the frame where

$$p = \left(P + \frac{M^2}{2P}, 0, 0, P\right)$$

and

$$q = \left(\frac{2M\nu - Q^2}{4P}, \mathbf{q}_\perp, -\frac{2M\nu + Q^2}{4P}\right),$$

with $P$ tending to infinity. Hence $Q^2 = q_\perp^2 + O(1/P^2)$ and $p \, . \, q = M\nu + O(1/P^2)$. Thus for terms leading in $P$ all constraints are satisfied, so that the frame may be regarded as appropriate for the implementation of parton ideas. We now assert that $0 < x_i < 1$, i.e. that all the partons move forward; this will be discussed critically in Chapter 2, but is assumed for the present. The $i$th parton of mass $\mu_i$ now has the four-momentum

$$p_i = \left(x_i P + \frac{\mu_i^2 + p_{i\perp}^2}{2x_i P}, \mathbf{p}_{i\perp}, x_i P\right).$$

Hence, in the frame under consideration, we may write

$$\sum_i E_i = E_\mathrm{p} + O\left(\frac{1}{P}\right),$$

and achieve leading energy conservation in the transition of the proton into partons. Moreover,

$$T \simeq \frac{2P}{\sum_i (p_{i\perp}^2+\mu_i^2)/x_i - M^2}, \qquad \tau \simeq \frac{4P}{2M\nu - Q^2}.$$

Hence the requirement $\tau \ll T$, for the validity of the parton model, may be expressed by means of the inequality

$$W^2 = M^2 + 2M\nu - Q^2 \gg m_\mathrm{eff}^2 - M^2,$$

where

$$m_\mathrm{eff}^2 = 2\sum_i \frac{p_{i\perp}^2+\mu_i^2}{x_i}.$$

Thus only when the mass of the inelastic state is large and *if the partons are effectively light with their transverse momenta kept under a finite and relatively small cut-off* can we use the parton idea in the present context. Moreover, if $\bar{\mathbf{p}}$ is the momentum (N.B. $\bar{p}_z = xP$) of the parton scattered by the virtual photon and $\bar{\mu}$ its mass, the energies of the parton before and after the said scattering are respectively the following:

$$\bar{p}_0 = xP + \frac{\bar{\mu}^2 + \bar{p}_\perp^2}{2xP},$$

$$\begin{aligned} \bar{p}_0' &= \sqrt{(\bar{\mathbf{p}}^2 + \bar{\mu}^2 + \mathbf{q}^2 + 2\bar{\mathbf{p}}\cdot\mathbf{q})} \\ &\simeq \left\{\left(xP - \frac{2M\nu+Q^2}{4P}\right)^2 + (\bar{\mathbf{p}}_\perp + \mathbf{q}_\perp)^2 + \bar{\mu}^2\right\}^{\frac{1}{2}} \\ &\simeq xP - \frac{2M\nu+Q^2}{4P} + \frac{\bar{p}_\perp^2 + \bar{\mu}^2 + 2\bar{\mathbf{p}}_\perp\cdot\mathbf{q}_\perp + Q^2}{2xP}. \end{aligned}$$

The energy difference across the vertex of transition from the partons (after scattering has taken place) to the final inelastic state $|n\rangle$ is

$$\Delta E' = \bar{p}_0 + q_0 - \bar{p}_0' - \sum_j{}' E_j = \frac{2M\nu - Q^2/x}{2P} - \frac{\bar{\mathbf{p}}_\perp\cdot\mathbf{q}_\perp}{xP} + \frac{M^2}{2P},$$

where $\sum_j'$ refers to the summation over all partons except the one scattered. For the instantaneity and incoherence of the scattering process to remain unaffected by final-state interactions, we also need the associated time period

$T' = 1/\Delta E'$ to be much greater than $\tau$. Comparing the corresponding expressions, we see that this is possible only if

$$x = \frac{Q^2}{2M\nu} \equiv w. \tag{1.2}$$

The consequent implication is that *the fraction of the longitudinal momentum carried by the scattered parton in the infinite momentum frame has to equal the scale variable* $w$. Moreover, with eqn (1.2), we may write

$$\bar{p}'_0 \simeq xP + \frac{\bar{\mu}^2 + \bar{p}_\perp^2}{2xP} + \frac{2M\nu - Q^2}{4P} + \frac{\bar{\mathbf{p}}_\perp \cdot \mathbf{q}_\perp}{xP} \simeq \bar{p}_0 + q_0 + O\left(\frac{\sqrt{Q^2}}{xP}\right). \tag{1.3}$$

In other words, to leading terms, four-momentum is conserved across the photon–parton vertex. Thus formally, in this situation, results from covariant perturbation theory may be used. In this chapter we shall employ that covariant formalism, with the understanding that the calculations ought really to be carried out in the infinite momentum frame and covariant results extracted therefrom. This procedure is the hallmark of kindergarten parton calculations. More proper theoretical approaches to the parton picture will be discussed in Chapter 2.

**Inelastic ep scattering**

We shall apply the above ideas to the problem of inelastic ep scattering by quantitatively setting up the equality of Fig. 15. Take the superposition of parton states to be $\sum_l a_l |l\rangle$, where $\{|l\rangle\}$ are a complete set of multi-particle parton states and $\sum_l |a_l|^2 = 1$. Then in the parton model, $|p\rangle \to \sum_l a_l |_l\rangle$ formally. Thus

$$W^{\mathrm{e}}_{\mu\nu} \equiv \frac{p_0}{M} \sum_n \langle p|J^{\mathrm{EM}}_\mu|n\rangle\langle n|J^{\mathrm{EM}}_\nu|p\rangle(2\pi)^6\delta^{(4)}(p+q-p_n)$$

$$\to \int_0^1 \mathrm{d}x \sum_l |a_l|^2 \sum_i \frac{p_0}{M} \int \mathrm{d}^3\bar{p}' \langle \bar{p}, i|j^{\mathrm{EM}}_\mu|\bar{p}', i\rangle\langle \bar{p}', i|j^{\mathrm{EM}}_\nu|\bar{p}, i\rangle \times$$

$$\times (2\pi^6)\delta^{(4)}(\bar{p}+q-\bar{p}')n_{li} f_{li}(x), \tag{1.4}$$

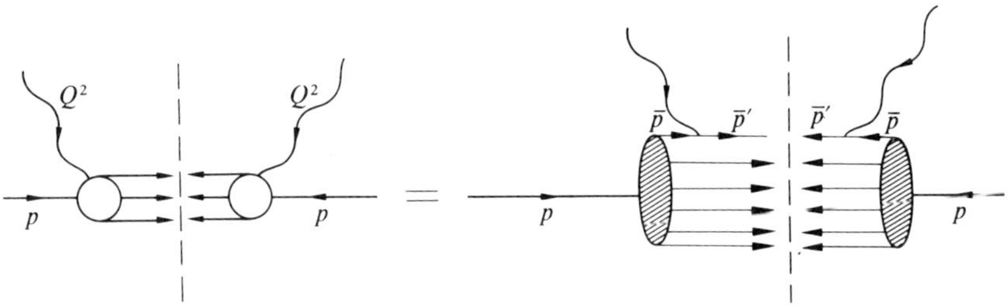


FIG. 15. Parton model of deep inelastic electron–proton scattering.

where $j_\mu$ is the bare one-body current operator (made up of free fields) acting with a point vertex, $f_{li}(x)$ is the probability density of finding a parton of type $i$ with longitudinal scale factor $x$ in the configuration $|l\rangle$, and $n_{li}$ is the number of partons of type $i$ in $|l\rangle$. The explicit calculations of $W^{e}_{\mu\nu}$, treating the partons first as spin-$\frac{1}{2}$ fermions and then as scalar bosons, will now be given.

*Spin-$\frac{1}{2}$ partons*

Substitute in eqn (1.4) the relation

$$\langle \bar{p}', i| j^{\mathrm{EM}}_{\nu} |\bar{p}, i\rangle = \left(\frac{1}{2\pi}\right)^3 \bar{u}(\bar{p}')\gamma_\nu u(\bar{p})\lambda_i \frac{\mu_i}{\sqrt{(\bar{p}_0\bar{p}'_0)}}, \tag{1.5}$$

where $\lambda_i$ is some multiple of $e$, carried as charge by the $i$th type of parton. For antipartons, we just replace $u$ by $v$. Since $\bar{p}_\mu \simeq xp_\mu$, we can write

$$W^{e}_{\mu\nu} \to \int_0^1 \mathrm{d}x \sum_{l,i} |a_l|^2 \frac{p_0}{M} \frac{\lambda_i^2}{\bar{p}_0\bar{p}'_0} \frac{1}{24} \mathrm{Tr}\{(\bar{p}' + \mu_i)\gamma_\nu(\bar{p}+\mu_i)\gamma_\mu\} \times$$
$$\times f_{li}(x) n_{li} \delta(\bar{p}_0 + q_0 - \bar{p}'_0)$$
$$= \int_0^1 \mathrm{d}x \sum_{l,i} |a_l|^2 \frac{1}{M} \lambda_i^2 f_{li}(x) n_{li} \left(2p_\mu p_\nu + \frac{p_\mu q_\nu + p_\nu q_\mu}{x} + \frac{q^2}{2x^2} g_{\mu\nu}\right) \times$$
$$\times \frac{x}{2M\nu}\delta\left(x - \frac{Q^2}{2M\nu}\right),$$

to leading terms. (Although here we treated only partons, the antiparton contribution term may be seen to be the same as above.) Then

$$\lim W^{e}_{\mu\nu} = \int_0^1 \mathrm{d}x \frac{w}{M^2\nu} \sum_l |a_l|^2 \sum_i f_{li}(x)\lambda_i^2 n_{li}\delta(x-w) \times$$
$$\times \left(2p_\mu p_\nu + \frac{p_\mu q_\nu + p_\nu q_\mu}{x} + \frac{q^2}{2x^2} g_{\mu\nu}\right) \tag{1.6}$$

to leading order. Note that the mass of the parton has disappeared from the leading terms. Now comparing eqn (1.6) with eqn (1.7) we have

$$\nu W^{e}_2 \to w \sum_{l,i} |a_l|^2 f_{li}(w)\lambda_i^2 n_{li} = F^{e}_2(w), \tag{1.7}$$

$$M W^{e}_1 \to \frac{1}{2} \sum_{l,i} |a_l|^2 f_{li}(w)\lambda_i^2 n_{li} = F^{e}_1(w). \tag{1.8}$$

Moreover,

$$\frac{MW^{e}_1}{\nu W^{e}_2} \to \frac{1}{2w},$$

so that from eqns (I.18) and (I.19) we have

$$\sigma_T \neq 0, \qquad \sigma_L \to 0, \qquad R \equiv \sigma_L/\sigma_T \to 0.$$

Finally, we can write, for spin-$\frac{1}{2}$ partons,

$$F_1^e(w) = \frac{1}{2w} F_2^e(w). \tag{1.9}$$

*Spin-0 partons*

Here

$$\langle \bar{p}', i | j_\nu^{EM} | \bar{p}, i \rangle = \frac{1}{(2\pi)^3} \frac{1}{\sqrt{(4\bar{p}_0 \bar{p}_0')}} (\bar{p} + \bar{p}')_\nu \lambda_i, \tag{1.10}$$

and

$$\lim W_{\mu\nu}^e = \int_0^1 dx \sum_{l,i} |a_l|^2 f_{li}(x) n_{li} \frac{p_0}{M} \frac{1}{4\bar{p}_0 \bar{p}_0'} \lambda_i^2 (2\bar{p}+q)_\mu (2\bar{p}+q)_\nu \times$$
$$\times \delta\left(\frac{2M\nu - (Q^2/x)}{P}\right)$$
$$= \int_0^1 dx \sum_{l,i} |a_l|^2 \lambda_i^2 \frac{n_{li}}{M} \left(2p_\mu p_\nu + \frac{p_\mu q_\nu + p_\nu q_\mu}{x} + \frac{q_\mu q_\nu}{x^2}\right) \frac{x}{2M\nu} \delta(x - w) \tag{1.11}$$

to leading terms. This gives

$$\nu W_2^e \to w \sum_l |a_l|^2 \sum_i f_{li}(w) \lambda_i^2 n_{li}, \tag{1.12a}$$

$$W_1^e \to 0, \tag{1.12b}$$

the latter following from the absence of any $g_{\mu\nu}$ term. Moreover, eqns (I.18) and (I.19) imply that in the present case

$$\sigma_L \neq 0, \qquad \sigma_T \to 0,$$

so that

$$R \equiv \frac{\sigma_L}{\sigma_T} \to \infty.$$

Experimentally, $R = 0{\cdot}18 \pm 0{\cdot}2$ (Kendall 1971), which is consistent with zero. So, even if there is a weighted sum of spin-$\frac{1}{2}$ and spin-0 partons contributing to $W_{\mu\nu}^e$, the amount of the latter contribution should be very small. The connection between the ratio $\sigma_L/\sigma_T$ and the spin of partons can be seen

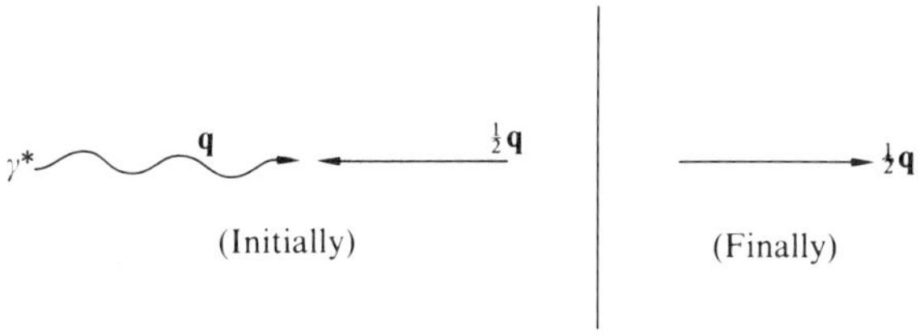


FIG. 16. Breit frame for linking $\sigma_L/\sigma_T$ with the parton spin.

physically by using helicity arguments in the Breit frame illustrated in Fig. 16 (see Exercise 1.1).

*Comments*

It is customary to define the probability and the average squared charge for the configuration $|l\rangle$ as $|a_l|^2 \equiv P(l)$ and $\sum_i n_{li}\lambda_i^2 = \langle \sum_i \lambda_i^2 \rangle_l$ respectively. Thus, for example, we can write

$$\lim \nu W_2^e = w \sum_l P(l) f_l(w) \left\langle \sum_i \lambda_i^2 \right\rangle_l . \tag{1.13}$$

The experimental scaling curves are shown in Fig. 17 (Miller *et al.* 1972). Note that $F_2$ vanishes at $\omega = 1$ which corresponds to the threshold.

## Inelastic ($\nu$, $\bar{\nu}$) p scattering

Using parton ideas, we have obtained the scaling behaviour of electromagnetic structure functions in the deep inelastic scattering of electrons from protons. Now we shall apply the same idea to the case of inelastic neutrino scattering. In the Introduction (p. 7), the differential cross-section for such a process has been given via tensors such as

$$W_{\mu\nu}^{\nu p} = \frac{1}{2\pi} \int d^4x \, e^{iq.x} \langle p | J_\mu^{W\dagger}(x) J_\nu^W(0) | p \rangle (2\pi)^3 \frac{p_0}{M} \tag{1.14}$$

and

$$W_{\mu\nu}^{\bar{\nu} n} = \frac{1}{2\pi} \int d^4x \, e^{iq.x} \langle n | J_\mu^W(x) J_\nu^{W\dagger}(0) | n \rangle (2\pi)^3 \frac{p_0}{M} . \tag{1.15}$$

Here $p$ and n refer to the two relevant target nucleons. $J_\mu^W$ is the hadronic part of the weak current and, further, has two components: (1) strangeness conserving and (2) strangeness changing. The contributions from the latter to quantities of present interest are proportional to the sine-squared of the Cabbibo angle $\theta_C$. Experimentally, $\sin^2\theta_C \simeq 0{\cdot}05$, so that for our

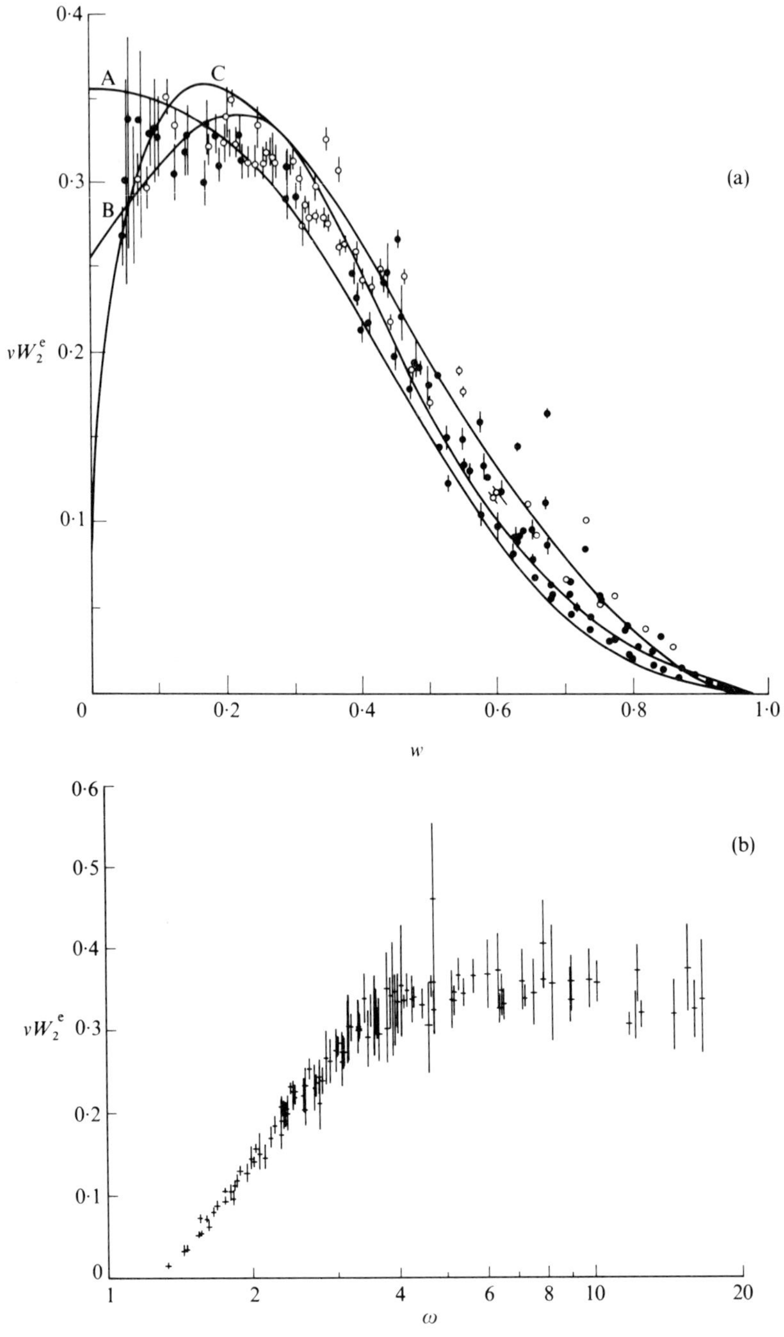


FIG. 17. (a) $\nu W_2^e$ as a function of $w$. (b) $\nu W_2^e$ as a function of $\omega$ (Miller *et al.* 1972).

purposes we may put $\sin\theta_C = 0$ and consider the hadronic weak vector current to be strangeness-conserving. Thus eqns (1.14) and (1.15) become (with 1, 2 being SU(2) indices)

$$W_{\mu\nu}^{\nu p} = \frac{p_0}{M}(2\pi)^2 \int d^4x\, e^{iq.x}\langle p|J_\mu^{1-i2}(x)J_\nu^{1+i2}(0)|p\rangle, \tag{1.16}$$

$$W_{\mu\nu}^{\bar{\nu}n} = \frac{p_0}{M}(2\pi)^2 \int d^4x\, e^{iq.x}\langle n|J_\mu^{1+i2}(x)J_\nu^{1-i2}(0)|n\rangle. \tag{1.17}$$

We can show from eqns (1.16) and (1.17) and isospin invariance that

$$W_{\mu\nu}^{\nu p} = W_{\mu\nu}^{\bar{\nu}n}, \qquad W_{\mu\nu}^{\bar{\nu}p} = W_{\mu\nu}^{\nu n} \tag{1.18}$$

(see Exercise 1.2). In view of the results (1.18) we can restrict our consideration to the $\nu$p and $\bar{\nu}$p cases only.

Once again, as in the case of electromagnetic probes, we take the point of view of Fig. 18, which depicts deep inelastic $\nu$p scattering.

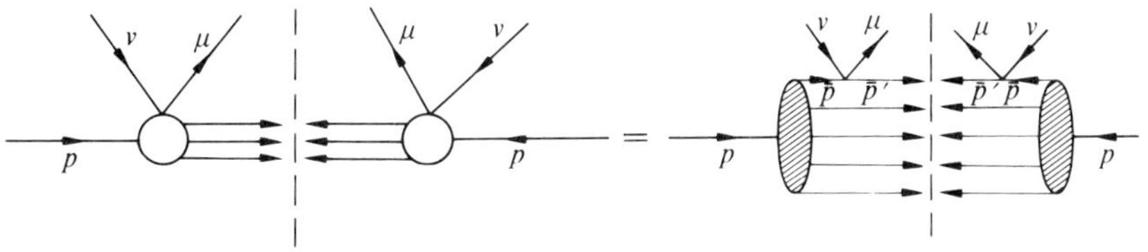


FIG. 18. Parton model of deep inelastic neutrino–proton scattering (cf. Fig. 15).

Note that in this case the weak current acting at the parton–neutrino vertex changes the charge of the parton by $\Delta Q = +1$. So we can write

$$\begin{aligned} W_{\alpha\beta}^{\nu} &= \frac{p_0}{M}\sum_n \langle p|J_\alpha^{W\dagger}|n\rangle\langle n|J_\beta^{W}|p\rangle(2\pi)^6\delta^{(4)}(p+q-p_n) \\ &\to \int_0^1 dx \sum_{l,i,i'} |a_l|^2 \frac{p_0}{M}\int d^3\bar{p}'\langle\bar{p},i|j_\alpha^{W\dagger}|\bar{p}',i'\rangle\langle\bar{p}',i'|j_\beta^{W}|\bar{p},i\rangle \times \\ &\quad \times (2\pi)^6\delta^{(4)}(\bar{p}+q-\bar{p}')n_{li}f_{li}(x), \end{aligned} \tag{1.19}$$

and similarly

$$\begin{aligned} W_{\alpha\beta}^{\bar{\nu}} &= \frac{p_0}{M}\sum_n \langle p|J_\alpha^{W}|n\rangle\langle n|J_\beta^{W\dagger}|p\rangle(2\pi)^6\delta^{(4)}(p+q-p_n) \\ &\to \int_0^1 dx \sum_{l,i,i'} |a_l|^2 \frac{p_0}{M}\int d^3\bar{p}'\langle\bar{p},i|j_\alpha^{W}|\bar{p}',i'\rangle\langle\bar{p}',i'|j_\beta^{W\dagger}|\bar{p},i\rangle \times \\ &\quad \times (2\pi)^6\delta^{(4)}(\bar{p}+q-\bar{p}')n_{li}f_{li}(x). \end{aligned} \tag{1.20}$$

Once again, $j_\mu$ is the one-body current operator made up of free fields acting with a point vertex.

Let us treat the partons as spin-$\frac{1}{2}$ fermions. Then we have

$$\langle \bar{p}', i' | j_\beta^{\mathrm{W}} | \bar{p}, i \rangle = \begin{cases} \left(\dfrac{1}{2\pi}\right)^3 \dfrac{\mu_i}{\sqrt{(\bar{p}_0 \bar{p}_0')}} \bar{u}_{i'}(\bar{p}')\gamma_\beta(1-\gamma_5)u_i(\bar{p})\lambda_i^{\mathrm{W}} (\text{partons}), \\ \left(\dfrac{1}{2\pi}\right)^3 \dfrac{\mu_i}{\sqrt{(\bar{p}_0 \bar{p}_0')}} \bar{v}_{i'}(\bar{p}')\gamma_\beta(1-\gamma_5)v_i(\bar{p})\lambda_i^{\mathrm{W}} (\text{antipartons}). \end{cases}$$

Here $\lambda_i^{\mathrm{W}}$ is a weak charge. It is unity if the parton of type $i$ belongs to an isospin multiplet and can be raised by an isospin raising operator to $i'$, but otherwise it vanishes. In the case of quarks it is unity for $q_n$ and $q_{\bar{p}}$, but zero for others. Substituting these expressions in eqns (1.19) and (1.20) and comparing coefficients from eqn (I.27) we have

$$\lim_{\mathrm{Bj}} MW_1^{\nu} = F_1^{\nu}(w) = \sum_l \sum_i f_{li}(w)|a_l|^2(\lambda_i^{\mathrm{W}})^2 n_{li}, \tag{1.21a}$$

$$\lim_{\mathrm{Bj}} \nu W_2^{\nu} = F_2^{\nu}(w) = 2w \sum_l \sum_i f_{li}(w)|a_l|^2(\lambda_i^{\mathrm{W}})^2 n_{li}, \tag{1.21b}$$

$$\lim_{\mathrm{Bj}} \nu W_3^{\nu} = F_3^{\nu}(w) = 2 \sum_l \sum_i f_{li}(w)|a_l|^2(\lambda_i^{\mathrm{W}})^2 n_{li}\eta_i, \tag{1.21c}$$

where the signature $\eta_i$ is negative for partons and positive for antipartons (i.e. when partons are antiparticles). Comparison of these forms with the corresponding ones in the inelastic electron-scattering case treated earlier shows that $F_{1,2}^{\nu}$ and $F_{1,2}^{\mathrm{e}}$ differ by an over-all factor of 2, but the forms are the same. The origin of this factor of 2 can be traced to the term $(1-\gamma_5)$ in the neutrino–parton interaction. While taking the trace this gives a factor $(1-\gamma_5)^2 = 2(1-\gamma_5)$. Note furthermore that, when we evaluate the trace in $W_{\mu\nu}$, the contributions of partons and antipartons, although the same for $W_{1,2}^{\nu}$, will differ in sign for $W_3^{\nu}$, because in the parton trace coming from $\sum_{\text{spins}} \langle j^{\mathrm{W}} \rangle^2$ the $\varepsilon^{\alpha\beta\gamma\delta}$ term flips sign on going from $u$-spinors to $v$-spinors.

For antineutrinos

$$\langle \bar{p}', i | j_\beta^{\mathrm{W}\dagger} | \bar{p}, i \rangle = \begin{cases} \left(\dfrac{1}{2\pi}\right)^3 \dfrac{\mu_i}{\sqrt{(\bar{p}_0 \bar{p}_0')}} \bar{u}(\bar{p}')\gamma_\beta(1-\gamma_5)u(\bar{p})\bar{\lambda}_i^{\mathrm{W}} \ (\text{partons}), \\ \left(\dfrac{1}{2\pi}\right)^3 \dfrac{\mu_i}{\sqrt{(\bar{p}_0 \bar{p}_0')}} \bar{v}(\bar{p}')\gamma_\beta(1-\gamma_5)v(\bar{p})\bar{\lambda}_i^{\mathrm{W}} \ (\text{antipartons}). \end{cases}$$

Here the weak charge $\bar{\lambda}_i^{\mathrm{W}}$ is unity if the parton $i$ can be lowered by an isospin-lowering operator to $i'$, but vanishes otherwise. Considering quarks again, the weak charge is unity for $q_p$ and $q_{\bar{n}}$ and zero for others. From this it follows, in the same way as above, that (see also Exercise 1.3)

$$F_1^{\bar{\nu}}(w) = \sum_l \sum_i f_{li}(w)|a_l|^2(\bar{\lambda}_i^{\mathrm{W}})^2 n_{li}, \tag{1.22a}$$

$$F_2^{\bar{\nu}}(w) = 2w \sum_l \sum_i f_{li}(w)|a_l|^2(\bar{\lambda}_i^{\mathrm{W}})^2 n_{li}, \tag{1.22b}$$

$$F_3^{\bar{\nu}}(w) = 2 \sum_l \sum_i f_{li}(w)|a_l|^2(\bar{\lambda}_i^{\mathrm{W}})^2 n_{li}\eta_i. \tag{1.22c}$$

Thus

$$F_1^{\nu\bar{\nu}} = \frac{1}{2w} F_2^{\nu\bar{\nu}}, \tag{1.23a}$$

$$F_3^{\nu} = \mp 2F_1^{\nu} = \mp \frac{1}{w} F_2^{\nu}, \tag{1.23b}$$

$$F_3^{\bar{\nu}} = \mp 2F_1^{\bar{\nu}} = \mp \frac{1}{w} F_2^{\bar{\nu}}, \tag{1.23c}$$

where the upper and lower signs refer to parton and antiparton contributions respectively.

If we define $F_{2,3\mathscr{P}}^{\nu\bar{\nu}}$ and $F_{2,3\bar{\mathscr{P}}}^{\nu\bar{\nu}}$ to be the contributions to those functions from partons and antipartons respectively, we can write

$$F_{3\mathscr{P}}^{\nu\bar{\nu}} = -\frac{1}{w} F_{2\mathscr{P}}^{\nu\bar{\nu}} \quad \text{and} \quad F_{3\bar{\mathscr{P}}}^{\nu\bar{\nu}} = \frac{1}{w} F_{2\bar{\mathscr{P}}}^{\nu\bar{\nu}}. \tag{1.24}$$

Consider now the differential cross-section formula

$$\frac{\mathrm{d}^2\sigma^{(\nu\bar{\nu})}}{\mathrm{d}Q^2\,\mathrm{d}\nu} = \frac{G^2}{2\pi}\frac{p_{\mu0}}{p_{\nu0}}\left\{2W_1^{\nu\bar{\nu}}\sin^2\frac{\theta_\mu}{2} + W_2^{\nu\bar{\nu}}\cos^2\frac{\theta_\mu}{2} \right.$$
$$\left. \pm W_3^{\nu\bar{\nu}}\frac{p_{\mu0}+p_{\nu0}}{M}\sin^2\frac{\theta_\mu}{2}\right\}. \tag{1.25}$$

Let us introduce a new variable $y$, called inelasticity, by $y \equiv \nu/p_{\nu0} = (p_{\nu0} - p_{\mu0})/p_{\nu0}$. Clearly $0 < y < 1$. Then

$$Q^2 \simeq 4p_{\nu0}p_{\mu0}\sin^2\frac{\theta_\mu}{2} = 4\frac{\nu^2}{y^2}(1-y)\sin^2\frac{\theta_\mu}{2}$$

and, moreover,

$$\mathrm{d}Q^2\,\mathrm{d}\nu = 2\frac{M\nu^2}{y}\,\mathrm{d}y\,\mathrm{d}w.$$

In the scaling limit then, eqn (1.25) yields

$$\lim_{\substack{\nu\to\infty \\ w\ \text{fixed}}} \frac{\mathrm{d}^2\sigma^{(\nu\bar{\nu})}}{\mathrm{d}w\,\mathrm{d}y} = \frac{G^2Mp_{\nu0}}{\pi}\left\{(1-y)F_2^{\nu\bar{\nu}} + y^2wF_1^{\nu\bar{\nu}} \mp y\left(1-\frac{y}{2}\right)wF_3^{\nu\bar{\nu}}\right\}. \tag{1.26}$$

Using

$$F_1 = \frac{1}{2w}F_2 \quad \text{and} \quad F_{2,3} = F_{2,3\mathscr{P}} + F_{2,3\bar{\mathscr{P}}},$$

we obtain

$$\lim_{\substack{\nu\to\infty \\ w\,\text{fixed}}} \frac{\mathrm{d}^2\sigma^{(\nu)}}{\mathrm{d}w\,\mathrm{d}y} = \frac{G^2 M p_{\nu 0}}{\pi}\{F^{\nu}_{2\mathscr{P}}(w) + (1-y)^2 F^{\nu}_{2\bar{\mathscr{P}}}(w)\}, \tag{1.27a}$$

$$\lim_{\substack{\nu\to\infty \\ w\,\text{fixed}}} \frac{\mathrm{d}^2\sigma^{(\bar{\nu})}}{\mathrm{d}w\,\mathrm{d}y} = \frac{G^2 M p_{\bar{\nu} 0}}{\pi}\{(1-y)^2 F^{\bar{\nu}}_{2\mathscr{P}}(w) + F^{\bar{\nu}}_{2\bar{\mathscr{P}}}(w)\}. \tag{1.27b}$$

On integrating eqns (1.27), we have

$$\lim_{p_{\nu 0}\to\infty} \sigma^{(\nu)} = \frac{G^2 M p_{\nu 0}}{\pi}\int_0^1 \mathrm{d}w\,\{F^{\nu}_{2\mathscr{P}}(w) + \tfrac{1}{3}F^{\nu}_{2\bar{\mathscr{P}}}(w)\}, \tag{1.28a}$$

$$\lim_{p_{\bar{\nu} 0}\to\infty} \sigma^{(\bar{\nu})} = \frac{G^2 M p_{\bar{\nu} 0}}{\pi}\int_0^1 \mathrm{d}w\,\{\tfrac{1}{3}F^{\bar{\nu}}_{2\mathscr{P}}(w) + F^{\bar{\nu}}_{2\bar{\mathscr{P}}}(w)\}. \tag{1.28b}$$

In obtaining eqns (1.28) we have (unjustifiably?) discarded the low-$Q^2$ non-scaling contribution. However, experimental data (Fig. 19) up to 10 GeV for $p_{\nu 0}$—averaged over protons and neutrons—do suggest (Perkins 1972) a linear rise of $\sigma^{(\nu\bar{\nu})}$ with incident energy. A similar feature has recently been observed up to $p_{\nu 0} \simeq 150$ GeV (Benvenutti *et al.* 1973).

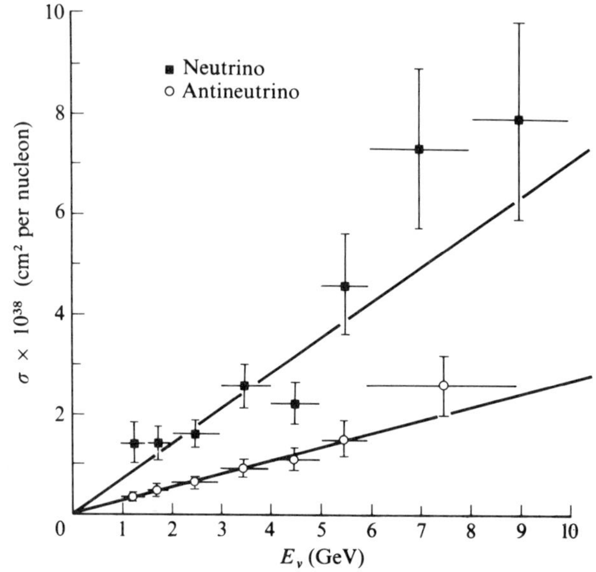


FIG. 19. Total neutrino and antineutrino cross-sections (Perkins 1972).

Since $(F^{\nu}_{2\mathscr{P}}+\frac{1}{3}F^{\nu}_{2\bar{\mathscr{P}}})$, $(\frac{1}{3}F^{\bar{\nu}}_{2\mathscr{P}}+F^{\bar{\nu}}_{2\bar{\mathscr{P}}}) \leqslant F^{\nu}_{2}, F^{\bar{\nu}}_{2}$, we obtain immediately the equality

$$\lim_{p_{\nu 0}\to\infty} \sigma^{(\nu\bar{\nu})} \leqslant \frac{G^2 M p_{\nu 0}}{\pi} \int_0^1 \mathrm{d}w\, F_2^{\nu\bar{\nu}}.$$

Experimentally (Perkins 1972),

$$\tfrac{1}{2}(\sigma^{(\nu\mathrm{p})}+\sigma^{(\nu\mathrm{n})}) \simeq \frac{G^2 M p_{\nu 0}}{\pi}(0{\cdot}52\pm 0{\cdot}13),$$

so that, with $\sigma^{(\nu\mathrm{n})} = \sigma^{(\bar{\nu}\mathrm{p})}$ and $F_2^{\nu\mathrm{n}} = F_2^{\bar{\nu}\mathrm{p}}$, we obtain

$$\int_0^1 \mathrm{d}w\,(F_2^{\nu\mathrm{p}}+F_2^{\bar{\nu}\mathrm{p}}) \geqslant 1{\cdot}08\pm 0{\cdot}27. \tag{1.29}$$

Moreover, if we make the *ad hoc* assumption that the antiparton contributions to $F_2^{\nu\bar{\nu}}$ are suppressed, eqns (1.28) lead to the result that, for neutrinos and antineutrinos with the same incident energy and averaged over protons and neutrons

$$\lim \sigma^{(\bar{\nu})}/\lim \sigma^{(\nu)} = \tfrac{1}{3}. \tag{1.30}$$

Amazingly enough, eqn (1.30) is in broad agreement with experiment—again averaged over protons and neutrons—over a sizeable energy range (Benvenutti *et al.* 1973) up to 150 GeV.

The case of spin-0 partons can be considered in a manner analogous to that shown above. In this case $F_1^{\nu,\bar{\nu}}$ vanishes (see also Exercise 1.4).

## Quark–partons

Our previous derivation of the scaling relations for electron and neutrino scattering did not need any special input, such as the identification of the partons, we used only the possible spin properties. We can go further and identify the partons with the quarks of Gell-Mann and Zweig (GMZ). This means treating the collection of partons as an assembly of $q_p$, $q_n$, and $q_\lambda$ quarks (with charges $\lambda = \frac{2}{3}, -\frac{1}{3}, -\frac{1}{3}$ respectively) and $q_{\bar{p}}$, $q_{\bar{n}}$, and $q_{\bar{\lambda}}$ antiquarks (with opposite charges). We are then led to specific sum rules and inequalities as consequences of various conservation laws (Llewellyn-Smith 1972*a*). In what follows we investigate these results. Throughout, we consider exact symmetries of strong interactions rather than approximate ones such as SU(3).

Let us assume that in place of partons we have three valence quarks plus a 'sea' of quark-antiquark pairs in an isosinglet state. The sea does not contribute to the over-all properties of the hadron, such as isospin, charge, and baryon number. Then from eqns (1.8) and (1.13) we have

$$F_1^{\mathrm{ep}} - F_1^{\mathrm{en}} = \frac{1}{2}\left[\frac{4}{9}\left\{\sum_l P^{\mathrm{p}}(l) f^{\mathrm{p}}_{l\mathrm{q_p}} n^{\mathrm{p}}_{l\mathrm{q_p}} - \sum_l P^{\mathrm{n}}(l) f^{\mathrm{n}}_{l\mathrm{q_p}} n^{\mathrm{n}}_{l\mathrm{q_p}}\right\} + \right.$$
$$\left. + \frac{1}{9}\left\{\sum_l P^{\mathrm{p}}(l) f^{\mathrm{p}}_{l\mathrm{q_n}} n^{\mathrm{p}}_{l\mathrm{q_n}} - \sum_l P^{\mathrm{n}}(l) f^{\mathrm{n}}_{l\mathrm{q_n}} n^{\mathrm{n}}_{l\mathrm{q_n}}\right\}\right]. \tag{1.31}$$

In eqn (1.31) the superscripts p and n refer to proton and neutron targets respectively. We note, moreover, that the contribution from the sea has been cancelled out by isospin invariance. We shall now use two further relations which follow from isospin invariance, indeed they follow simply from charge symmetry. These are

$$\sum_l P^{\mathrm{p}}(l) f^{\mathrm{p}}_{l\mathrm{q_p}} n^{\mathrm{p}}_{l\mathrm{q_p}} = \sum_l P^{\mathrm{n}}(l) f^{\mathrm{n}}_{l\mathrm{q_n}} n^{\mathrm{n}}_{l\mathrm{q_n}}, \tag{1.32a}$$

$$\sum_l P^{\mathrm{p}}(l) f^{\mathrm{p}}_{l\mathrm{q_n}} n^{\mathrm{p}}_{l\mathrm{q_n}} = \sum_l P^{\mathrm{n}}(l) f^{\mathrm{n}}_{l\mathrm{q_p}} n^{\mathrm{n}}_{l\mathrm{q_p}}. \tag{1.32b}$$

Thus we have

$$F_1^{\mathrm{ep}} - F_1^{\mathrm{en}} = \tfrac{1}{6}\sum_l P^{\mathrm{p}}(l)(f^{\mathrm{p}}_{l\mathrm{q_p}} n^{\mathrm{p}}_{\mathrm{q_p}} - f^{\mathrm{p}}_{l\mathrm{q_n}} n^{\mathrm{p}}_{\mathrm{q_n}}). \tag{1.33}$$

Similarly, we can easily obtain the result

$$F_3^{\nu\mathrm{p}} - F_3^{\bar{\nu}\mathrm{p}} = 2\sum_l P^{\mathrm{p}}(l)\,(f^{\mathrm{p}}_{l\mathrm{q_p}} n^{\mathrm{p}}_{l\mathrm{q_p}} - f^{\mathrm{p}}_{l\mathrm{q_n}} n^{\mathrm{p}}_{\mathrm{q_n}}). \tag{1.34}$$

Eqns (1.33) and (1.34) yield the equality (Llewellyn-Smith 1972*a*)

$$F_3^{\nu\mathrm{p}} - F_3^{\bar{\nu}\mathrm{p}} = 12(F_1^{\mathrm{ep}} - F_1^{\mathrm{en}}). \tag{1.35a}$$

Similar arguments, strengthened with positivity, lead to (see Exercise 1.5) (Llewellyn-Smith 1972*b*) the inequality

$$F_1^{\mathrm{ep}} + F_1^{\mathrm{en}} \geqslant \tfrac{5}{18}(F_1^{\nu\mathrm{p}} + F_1^{\bar{\nu}\mathrm{p}}). \tag{1.35b}$$

If we use the result from SLAC (Kendall 1971) that

$$\int_0^1 \mathrm{d}w\,\{F_2^{\mathrm{ep}}(w) + F_2^{\mathrm{en}}(w)\} \simeq 0{\cdot}28 \pm 0{\cdot}04,$$

and compare it with eqn (1.29), we see that the integrated Llewellyn-Smith inequality (with $F_2 = F_1/2w$) is almost saturated. Eqn (1.35b) does of course reduce to an equality if the contribution from the sea is summarily ignored.

*Gottfried sum rule*

A very simple sum rule follows from the normalization condition on the probability for the longitudinal fraction $x$, i.e. $\int_0^1 \mathrm{d}x\, f_{li}(x) = 1$ and from eqn (1.7), namely,

$$\int_0^1 \frac{\mathrm{d}w}{w} F_2^{\mathrm{e}}(w) = \sum_{l,i} P(l) n_{li} \lambda_i^2. \tag{1.36}$$

This is the Gottfried sum rule, which has the following implication. Although an experimental knowledge of $F_2^{\mathrm{ep}}(w)$ very near $w = 0$ is not available, the data (Fig. 17) suggest that $F_2^{\mathrm{ep}}(w = 0)$ is non-zero. The left-hand side of eqn (1.36) then develops a logarithmic divergence at the lower limit of the integral. In that case the sum rule implies that the number of charged constituents in the nucleon that contribute to $F_2^{\mathrm{ep}}(w)$ is infinite. Occasionally it is alleged that the domain near $w = 0$ is really outside the purview of parton considerations and belongs to the 'diffraction picture'. According to this reasoning, we should apply the parton analysis only to $F_2^{\mathrm{e}}(w) - F_2^{\mathrm{e}}(0)$, so that the Gottfried sum rule becomes

$$\int_0^1 \frac{\mathrm{d}w}{w} \{F_2^{\mathrm{e}}(w) - F_2^{\mathrm{e}}(0)\} \stackrel{?}{=} \sum_{l,i} P(l) n_{li} \lambda_i^2,$$

thus freed from the *bête noire* of the region of the logarithmic divergence. Since this involves a somewhat arbitrary procedure, we shall not elaborate on this question.

We now try to get results avoiding possible trouble of the above type at $w = 0$. Consider a proton target. Taking quark–partons as constituents and assuming that the sea of $\mathrm{q\bar{q}}$ pairs is uniform, we have

$$\begin{aligned}\sum_l P^{\mathrm{p}}(l) \sum_i n_{li}^{\mathrm{p}} \lambda_i^2 &= 2\left(\frac{2}{3}\right)^2 + \left(\frac{1}{3}\right)^2 + \sum_l \frac{1}{3}\left\{\left(\frac{2}{3}\right)^2 + \left(\frac{1}{3}\right)^2 + \left(\frac{1}{3}\right)^2\right\} P^{\mathrm{p}}(l)(l-3) \\ &= 1 + \tfrac{2}{9}(\langle l \rangle_{\mathrm{p}} - 3). \end{aligned} \tag{1.37}$$

For a neutron, the corresponding result is

$$\sum_l P^{\mathrm{n}}(l) \sum_i n_{li} \lambda_i^2 = \tfrac{2}{3} + \tfrac{2}{9}(\langle l \rangle_{\mathrm{n}} - 3). \tag{1.38}$$

Eqns (1.37) and (1.38) imply

$$\begin{aligned}\int_0^1 \frac{\mathrm{d}w}{w} \{F_2^{\mathrm{ep}}(w) - F_2^{\mathrm{en}}(w)\} &\equiv \int_1^\infty \frac{\mathrm{d}\omega}{\omega} \{F_2^{\mathrm{ep}}(\omega) - F_2^{\mathrm{en}}(\omega)\} \\ &= \tfrac{1}{3} + \tfrac{2}{9}\langle l \rangle_{\mathrm{p-n}}. \end{aligned} \tag{1.39}$$

The experimental data up to $\omega = 17$, if smoothly extrapolated to infinity, indicate (Miller *et al.* 1972) that

$$\int_1^\infty \frac{\mathrm{d}w}{w}\{F_2^{\mathrm{ep}}(w) - F_2^{\mathrm{en}}(w)\} \leqslant \tfrac{1}{3},$$

leading to the conclusion

$$\langle l\rangle_{\mathrm{p}} < \langle l\rangle_{\mathrm{n}}.$$

*CCSR and BCSR*

We now extend considerations similar to those made above to the scattering of neutrinos and antineutrinos. It is straightforward to obtain the result

$$\int_0^1 \frac{\mathrm{d}w}{w}\{F_2^{\bar{\nu}\mathrm{p}}(w) - F_2^{\nu\mathrm{p}}(w)\} = 2\sum_l P^{\mathrm{p}}(l)\{n^{\mathrm{p}}_{l\mathrm{q_p}} + n^{\mathrm{p}}_{l\mathrm{q}_{\bar{\mathrm{n}}}} - n^{\mathrm{p}}_{l\mathrm{q_n}} - n^{\mathrm{p}}_{l\mathrm{q}_{\bar{\mathrm{p}}}}\},$$

where $n^{\mathrm{p}}_{l\mathrm{q_p}}$ is the number of p-quarks in the state $|l\rangle$ corresponding to a target proton, and so on. Now, for any parton configuration $|l\rangle$, charge conservation demands the equality

$$1 = \tfrac{2}{3}n^{\mathrm{p}}_{l\mathrm{q_p}} - \tfrac{1}{3}n^{\mathrm{p}}_{l\mathrm{q_n}} - \tfrac{2}{3}n^{\mathrm{p}}_{l\mathrm{q}_{\bar{\mathrm{p}}}} + \tfrac{1}{3}n^{\mathrm{p}}_{l\mathrm{q}_{\bar{\mathrm{n}}}}.$$

Thus we have the charge-conservation sum rule (CCSR)

$$\int_0^1 \frac{\mathrm{d}w}{w}\{F_2^{\bar{\nu}\mathrm{p}}(w) - F_2^{\nu\mathrm{p}}(w)\} = 2. \tag{1.40}$$

A slightly different version of this result was originally obtained by Adler from current-algebraic methods (Adler and Dashen 1968). It is gratifying to see that the quark–parton model recovers the same as a consequence of charge conservation.

We now derive a similar sum rule based on the conservation of baryonic number and due originally to Gross and Llewellyn-Smith. Consider the relation

$$\int_0^1 \mathrm{d}w\,\{F_3^{\bar{\nu}\mathrm{p}}(w) + F_3^{\nu\mathrm{p}}(w)\} = 2\sum_l P^{\mathrm{p}}(l)(-n^{\mathrm{p}}_{l\mathrm{q_p}} + n^{\mathrm{p}}_{l\mathrm{q}_{\bar{\mathrm{n}}}} - n^{\mathrm{p}}_{l\mathrm{q_n}} + n^{\mathrm{p}}_{l\mathrm{q}_{\bar{\mathrm{p}}}}). \tag{1.41}$$

Now the law of the conservation of baryonic number implies that

$$n^{\mathrm{p}}_{l\mathrm{q_p}} + n^{\mathrm{p}}_{l\mathrm{q_n}} - n^{\mathrm{p}}_{l\mathrm{q_{\bar{n}}}} - n^{\mathrm{p}}_{l\mathrm{q_{\bar{p}}}} = 3.$$

Hence we have the baryon-number conservation sum rule (BCSR)

$$\int_0^1 \mathrm{d}w\,\{F_3^{\bar{\nu}\mathrm{p}}(w) + F_3^{\nu\mathrm{p}}(w)\} = -6. \tag{1.42}$$

*Momentum-conservation sum rules*

Crucial to the above parton picture is the idea that all the partons move forward, namely in the $z$-direction, in the infinite momentum frame ($P_z \to \infty$). In any configuration $|l\rangle$, the equality $1 = \int_0^1 \mathrm{d}x\,.\,x\sum_{l,i} n_{li} f_{li}(x)$ holds provided we sum over all charged partons and other neutral constituents which carry fractions of the longitudinal momentum $P$. But in the above discussions the summation over $i$ runs only over those partons 'seen' by the relevant currents $J_\mu^{\mathrm{EM}}$ and $J_\mu^{\mathrm{W}}$, $J_\mu^{\mathrm{W}\dagger}$. For example, the electromagnetic current $J_\mu^{\mathrm{EM}}$ does not see neutral partons, and hence these do not enter the corresponding summations of $n_{li} f_{li}(x)$ which appear in the electromagnetic case. Thus arises the following question: are there any other constituents (in the quark–parton model those other than quarks or antiquarks) which do not interact with the currents but carry a fraction of $P$? This brings us to the matter of gluons (this name, since they are supposed to glue the quarks in a hadron), which will be discussed below.

In the standard gluon model, which has been promoted by Gell-Mann (Gell-Mann and Nee'man 1964), the gluons are SU(3)-singlet vector bosons which glue the quarks by means of the SU(3)-invariant interaction $\mathscr{L}_{\mathrm{I}} = g\bar{q}\gamma^\mu q B_\mu$, $B_\mu$ being the field of the gluon. Thus, in general, momentum conservation implies the constraint

$$1 = \int_0^1 \mathrm{d}x\, x \sum_l \Big\{\sum_i n_{li} f_{li}(x) + \sum_j \varepsilon_{lj}\Big\}, \tag{1.43}$$

where $0 < \varepsilon_{lj} < 1$ and $\varepsilon_{lj}$ = fraction of $P$ carried by type $j$ of those partons in $|l\rangle$ not seen by the relevant current. Let us put $\sum_l P^{\mathrm{N}}(l)\varepsilon_{lj} = \varepsilon_j^{\mathrm{N}}$. Then from eqns (1.21) and (1.22) we have

$$\int_0^1 \mathrm{d}w\,\{F_2^{\nu\mathrm{p}}(w) + F_2^{\bar{\nu}\mathrm{p}}(w)\}$$

$$= 2\sum_l P^{\mathrm{p}}(l)\int_0^1 \mathrm{d}x\, x\{f^{\mathrm{p}}_{l\mathrm{q_{\bar{p}}}}(x) n^{\mathrm{p}}_{l\mathrm{q_{\bar{p}}}} + f^{\mathrm{p}}_{l\mathrm{q_n}}(x) n^{\mathrm{p}}_{l\mathrm{q_n}} + f^{\mathrm{p}}_{l\mathrm{q_p}}(x) n^{\mathrm{p}}_{l\mathrm{q_p}} + f^{\mathrm{p}}_{l\mathrm{q_{\bar{n}}}}(x) n^{\mathrm{p}}_{l\mathrm{q_{\bar{n}}}}\}, \tag{1.44}$$

or, with the subscript g standing for gluons,

$$\int_0^1 \mathrm{d}w\, F_2^{(\nu+\bar{\nu})\mathrm{p}}(w) = 2\sum_l P^{\mathrm{p}}(l)(1-\varepsilon^{\mathrm{p}}_{l\mathrm{q}\lambda}-\varepsilon^{\mathrm{p}}_{l\mathrm{q}\bar{\lambda}}-\varepsilon^{\mathrm{p}}_{l\mathrm{g}})$$

$$= 2(1-\varepsilon^{\mathrm{p}}_{\mathrm{q}\lambda}-\varepsilon^{\mathrm{p}}_{\mathrm{q}\bar{\lambda}}-\varepsilon^{\mathrm{p}}_{\mathrm{g}}) \leqslant 2.$$

But this inequality is not useful as it is, since we already have from the Llewellyn-Smith inequality eqn (1.35b) that

$$\int_0^1 \mathrm{d}w\,\{F_2^{\nu\mathrm{p}}(w)+F_2^{\bar{\nu}\mathrm{p}}(w)\} \leqslant \tfrac{18}{5}\int_0^1 \mathrm{d}w\,\{F_2^{\mathrm{ep}}(w)+F_2^{\mathrm{en}}(w)\} \simeq 0{\cdot}1,$$

the 0·1 coming from experiment (e.g. Kendall 1971). There is, however, one interesting consequence of eqn (1.43), as shown below. We have

$$\int_0^1 \mathrm{d}w\, F_2^{\mathrm{ep}}(w) = \sum_l P^{\mathrm{p}}(l)\int_0^1 \mathrm{d}x\; x[\tfrac{4}{9}\{f^{\mathrm{p}}_{l\mathrm{q_p}}(x)n^{\mathrm{p}}_{l\mathrm{q_p}}+f_{l\mathrm{q}_{\bar{\mathrm{p}}}}(x)n_{l\mathrm{q}_{\bar{\mathrm{p}}}}\}+$$

$$+\tfrac{1}{9}\{f^{\mathrm{p}}_{l\mathrm{q_n}}(x)n^{\mathrm{p}}_{l\mathrm{q_n}}+f^{\mathrm{p}}_{l\mathrm{q}_{\bar{\mathrm{n}}}}(x)n^{\mathrm{p}}_{l\mathrm{q}_{\bar{\mathrm{n}}}}\}+$$

$$+\tfrac{1}{9}\{f^{\mathrm{p}}_{l\mathrm{q}\lambda}(x)n^{\mathrm{p}}_{l\mathrm{q}\lambda}+f^{\mathrm{p}}_{l\mathrm{q}\bar{\lambda}}(x)n^{\mathrm{p}}_{l\mathrm{q}\bar{\lambda}}\}] \tag{1.45}$$

and

$$\int_0^1 \mathrm{d}w\, F_2^{\mathrm{en}}(w) = \sum_l P^{\mathrm{n}}(l)\int_0^1 \mathrm{d}x\,.\,x[\tfrac{4}{9}\{f^{\mathrm{n}}_{l\mathrm{q_p}}(x)n^{\mathrm{n}}_{l\mathrm{q_p}}+f^{\mathrm{n}}_{l\mathrm{q}_{\bar{\mathrm{p}}}}(x)n^{\mathrm{n}}_{l\mathrm{q}_{\bar{\mathrm{p}}}}\}+$$

$$+\tfrac{1}{9}\{f^{\mathrm{n}}_{l\mathrm{q_n}}(x)n^{\mathrm{n}}_{l\mathrm{q_n}}+f^{\mathrm{n}}_{l\mathrm{q}_{\bar{\mathrm{n}}}}(x)n^{\mathrm{n}}_{l\mathrm{q}_{\bar{\mathrm{n}}}}\}+$$

$$+\tfrac{1}{9}\{f^{\mathrm{n}}_{l\mathrm{q}\lambda}(x)n^{\mathrm{n}}_{l\mathrm{q}\lambda}+f^{\mathrm{n}}_{l\mathrm{q}\bar{\lambda}}(x)n^{\mathrm{n}}_{l\mathrm{q}\bar{\lambda}}\}]. \tag{1.46}$$

Isospin invariance implies eqns (1.32), as well as the relation

$$\sum_l P^{\mathrm{p}}(l)f^{\mathrm{p}}_{l\mathrm{q}\lambda}n^{\mathrm{p}}_{l\mathrm{q}\lambda} = \sum_l P^{\mathrm{n}}(l)f^{\mathrm{n}}_{l\mathrm{q}\bar{\lambda}}n^{\mathrm{n}}_{l\mathrm{q}\bar{\lambda}} \tag{1.47}$$

and corresponding relations for the antiquarks. Substituting eqns (1.32), (1.47), and the corresponding antiquark relations in eqns (1.45) and (1.46), we have

$$\int_0^1 \mathrm{d}w\,\{F_2^{\mathrm{ep}}(w)+F_2^{\mathrm{en}}(w)\} = \sum_l P^{\mathrm{p}}(l)\int_0^1 \mathrm{d}x\; x[\tfrac{5}{9}\{f^{\mathrm{p}}_{l\mathrm{q_p}}(x)n^{\mathrm{p}}_{l\mathrm{q_p}}+f^{\mathrm{p}}_{l\mathrm{q}_{\bar{\mathrm{p}}}}(x)n^{\mathrm{p}}_{l\mathrm{q}_{\bar{\mathrm{p}}}}\}+$$

$$+\tfrac{5}{9}\{f^{\mathrm{p}}_{l\mathrm{q_n}}(x)n^{\mathrm{p}}_{l\mathrm{q_n}}+f^{\mathrm{p}}_{l\mathrm{q}_{\bar{\mathrm{n}}}}(x)n^{\mathrm{p}}_{l\mathrm{q}_{\bar{\mathrm{n}}}}\}+$$

$$+\tfrac{2}{9}\{f^{\mathrm{p}}_{l\mathrm{q}\lambda}(x)n^{\mathrm{p}}_{l\mathrm{q}\lambda}+f^{\mathrm{p}}_{l\mathrm{q}\bar{\lambda}}(x)n^{\mathrm{p}}_{l\mathrm{q}\bar{\lambda}}\}]. \tag{1.48}$$

Using eqns (1.44), (1.48), and our identity $F_2^{\bar{\nu}\mathrm{p}} = F_2^{\nu\mathrm{n}}$, we can write

$$\int_0^1 \mathrm{d}w[\tfrac{3}{4}\{F_2^{\nu\mathrm{p}}(w)+F_2^{\nu\mathrm{n}}(w)\}-\tfrac{9}{2}\{F_2^{\mathrm{ep}}(w)+F_2^{\mathrm{en}}(w)\}]$$

$$= -\sum_l P^{\mathrm{p}}(l)\int_0^1 \mathrm{d}x\, x\{f^{\mathrm{p}}_{l\mathrm{q_p}}(x)n^{\mathrm{p}}_{l\mathrm{q_p}}+f^{\mathrm{p}}_{l\mathrm{q}_{\bar{\mathrm{p}}}}(x)n^{\mathrm{p}}_{l\mathrm{q}_{\bar{\mathrm{p}}}}+f^{\mathrm{p}}_{l\mathrm{q_n}}(x)n^{\mathrm{p}}_{l\mathrm{q_n}}+f^{\mathrm{p}}_{l\mathrm{q}_{\bar{\mathrm{n}}}}(x)n^{\mathrm{p}}_{l\mathrm{q}_{\bar{\mathrm{n}}}}+$$

$$+f^{\mathrm{p}}_{l\mathrm{q}_\lambda}(x)n^{\mathrm{p}}_{l\mathrm{q}_\lambda}+f^{\mathrm{p}}_{l\mathrm{q}_{\bar{\lambda}}}(x)n^{\mathrm{p}}_{l\mathrm{q}_{\bar{\lambda}}}\}$$

$$= -1+\varepsilon_{\mathrm{g}}^{\mathrm{p}}.$$

Thus we have the sum rule

$$\varepsilon_{\mathrm{g}}^{\mathrm{p}} = 1+\int_0^1 \mathrm{d}w[\tfrac{3}{4}\{F_2^{\nu\mathrm{p}}(w)+F_2^{\nu\mathrm{n}}(w)\}-\tfrac{9}{2}\{F_2^{\mathrm{ep}}(w)+F_2^{\mathrm{en}}(w)\}]. \tag{1.49}$$

Present experimental figures are (Perkins 1972; Kendall 1971)

$$\int_0^1 \mathrm{d}w\{F_2^{\nu\mathrm{p}}(w)+F_2^{\nu\mathrm{n}}(w)\} \geqslant 1{\cdot}08 \pm 0{\cdot}27, \tag{1.50a}$$

$$\int_0^1 \mathrm{d}w\,\{F_2^{\mathrm{ep}}(w)+F_2^{\mathrm{en}}(w)\} \simeq 0{\cdot}28 \pm 0{\cdot}004. \tag{1.50b}$$

Hence we see that

$$\varepsilon_{\mathrm{g}} \geqslant 0{\cdot}52 \pm 0{\cdot}38, \tag{1.51}$$

i.e that the contribution from gluons to the proton's longitudinal momentum in the infinite momentum frame is nearly 50 per cent. Thus the requirement of gluons in the quark–parton model is unambiguously shown by experiment, through eqn (1.51). We note here that this is a general result in the quark–parton picture, as a consequence of momentum conservation. To obtain further results we need stronger assumptions, as discussed below.

We now make the assumption of *symmetric momentum distribution.* A rationale for this may be given (Bjorken and Paschos 1969), but we shall not go into such details here. We state it as an equation which speaks for itself,

$$\int_0^1 \mathrm{d}x\,.\,xf_{li}(x) = 1/l \tag{1.52}$$

for any $i$. We can discuss two cases.

1. *With no gluons*, we obtain

$$\int_0^1 \mathrm{d}w\, F_2^{\mathrm{ep}}(w) = \sum_l P^{\mathrm{p}}(l)\frac{1}{l}\sum_i n_{li}^{\mathrm{p}}\lambda_i^2. \tag{1.53}$$

Let us now adopt the picture of three core quarks plus a uniform sea of $\bar{\mathrm{q}}\mathrm{q}$ pairs, introduced earlier. Then we have

$$\int_0^1 \mathrm{d}w\, F_2^{\mathrm{ep}}(w) = \sum_l P^{\mathrm{p}}(l)\frac{1}{l}\left\{2\,.\,\frac{4}{9}+\frac{1}{9}-\frac{l-3}{3}\left(\frac{4}{9}+\frac{1}{9}+\frac{1}{9}\right)\right\}$$

$$= \frac{2}{9}+\frac{1}{3}\left\langle\frac{1}{l}\right\rangle \geqslant 0{\cdot}22. \tag{1.54}$$

But the left-hand side of eqn (1.54) is experimentally (Taylor 1969) near 0·17. Similarly for the neutron, we have

$$\int_0^1 \mathrm{d}w\, F_2^{\mathrm{en}}(w) = \sum_l P^{\mathrm{n}}(l)\frac{1}{l}\left\{\frac{4}{9}+2\,.\,\frac{1}{9}+\frac{l-3}{3}\left(\frac{4}{9}+\frac{1}{9}+\frac{1}{9}\right)\right\}$$

$$= \tfrac{2}{9} \simeq 0{\cdot}22, \tag{1.55}$$

where the left-hand side is experimentally near 0·12 (Kendall 1971). These instances of disagreement with experiment also demand the presence of gluons, unless we disbelieve in symmetric momentum distribution.

2. *With gluons*, we obtain (noting that in eqn (1.39) also $\langle l\rangle$ now is replaced by $\langle l-l_{\mathrm{g}}\rangle$ in the numerator) the following expressions for the same integrals,

$$\int_0^1 \mathrm{d}w\, F_2^{\mathrm{ep}}(w) = \frac{2}{9}\left(1-\left\langle\frac{l_{\mathrm{g}}}{l}\right\rangle\right)+\frac{1}{3}\left\langle\frac{1}{l}\right\rangle, \tag{1.56}$$

$$\int_0^1 \mathrm{d}w\, F_2^{\mathrm{en}}(w) = \frac{2}{9}\left(1-\left\langle\frac{l_{\mathrm{g}}}{l}\right\rangle\right). \tag{1.57}$$

With the above experimental numbers, we are led to conclude that

$$\left\langle\frac{l_{\mathrm{g}}}{l}\right\rangle_{\mathrm{p}} \simeq \frac{3}{2}\left\langle\frac{1}{l}\right\rangle+\frac{9}{40} \geqslant \frac{9}{40}, \tag{1.58}$$

$$\left\langle\frac{l_{\mathrm{g}}}{l}\right\rangle_{\mathrm{n}} \simeq \frac{9}{20}. \tag{1.59}$$

Note that the isoscalar nature of the gluons implies $\langle l_g/l\rangle_p = \langle l_g/l\rangle_n = \frac{9}{20}$. Since

$$\varepsilon_g = \sum_l P(l)\varepsilon_{l_g} = \sum_l P(l)\int_0^1 dx\,.\,x\sum_g f_{l_g}n_{l_g}$$
$$= \sum_l P(l)\frac{1}{l}n_{l_g} = \left\langle\frac{l_g}{l}\right\rangle,$$

we conclude that $\langle l_g/l\rangle \simeq 0{\cdot}5$, which is not inconsistent with eqn (1.51). Further, eqn (1.58) now implies that

$$\langle 1/l\rangle_p \simeq \tfrac{3}{20}.$$

Thus a nice physical picture emerges, giving an idea of the gluon admixture in a nucleon, namely, that nearly half of the partons are gluons. Moreover, on the average, the virtual photon 'sees' a configuration of only about 7 partons. The smallness of this number clearly has some significance, but it is not fully understood at the moment.

*Other isospin inequalities*

These may be derived from positivity, using the full content of isospin invariance (Nachtmann 1971). First let us simplify the notation by defining

$$u^p_{q_p} \equiv \sum_l P^p(l)f^p_{lq_p}(w)n^p_{lq_p}, \quad \text{etc.} \tag{1.61}$$

These are positive by construction. Isospin invariance decrees:

$$u^p_{q_p} = u^n_{q_n}, \qquad u^p_{q_n} = u^n_{q_p}, \qquad u^p_{q_\lambda} = u^n_{q_\lambda}$$

and corresponding equalities for the $u_{\bar{q}}$s. Notice that the isospin content of the statement $u^p_{q_p} = u^n_{q_n}$ is analogous to that of the equality $\langle p|q_p^\dagger q_p|p\rangle = \langle n|q_n^\dagger q_n|n\rangle$, where the qs stand for the corresponding quark fields. The states $|p\rangle$ and $|n\rangle$ change under isospin rotation as

$$\exp(i\pi I_2)\begin{bmatrix}|p\rangle\\|n\rangle\end{bmatrix} = \begin{bmatrix}|n\rangle\\-|p\rangle\end{bmatrix},$$

where $I_2$ is the appropriate isospin generator. Similarly

$$\exp(i\pi I_2)\begin{bmatrix}|q_p\rangle\\|q_n\rangle\end{bmatrix} = \begin{bmatrix}|q\rangle\\-|q_p\rangle\end{bmatrix},$$

and so on. Now, in the quark–parton model, we can write

$$\frac{F_2^{en}(w)}{F_2^{ep}(w)} = \frac{\frac{4}{9}(u^n_{q_p}+u^n_{q_{\bar{p}}})+\frac{1}{9}(u^n_{q_n}+u^n_{q_{\bar{n}}})+\frac{1}{9}(u^n_{q_\lambda}+u^n_{q_{\bar{\lambda}}})}{\frac{4}{9}(u^p_{q_p}+u^p_{q_{\bar{p}}})+\frac{1}{9}(u^p_{q_n}+u^p_{q_{\bar{n}}})+\frac{1}{9}(u^p_{q_\lambda}+u^p_{q_{\bar{\lambda}}})}. \tag{1.62}$$

Imposing isospin invariance, we have

$$\frac{F_2^{\mathrm{en}}(w)}{F_2^{\mathrm{ep}}(w)} = \frac{u_{\mathrm{q_p}}^{\mathrm{p}} + u_{\mathrm{q_{\bar{p}}}}^{\mathrm{p}} + 4(u_{\mathrm{q_n}}^{\mathrm{p}} + u_{\mathrm{q_{\bar{n}}}}^{\mathrm{p}}) + u_{\mathrm{q_\lambda}}^{\mathrm{p}} + u_{\mathrm{q_{\bar{\lambda}}}}^{\mathrm{p}}}{4(u_{\mathrm{q_p}}^{\mathrm{p}} + u_{\mathrm{q_{\bar{p}}}}^{\mathrm{p}}) + u_{\mathrm{q_n}}^{\mathrm{p}} + u_{\mathrm{q_{\bar{n}}}}^{\mathrm{p}} + u_{\mathrm{q_\lambda}}^{\mathrm{p}} + u_{\mathrm{q_{\bar{\lambda}}}}^{\mathrm{p}}}. \tag{1.63}$$

Thus it follows that

$$\tfrac{1}{4} \leqslant F_2^{\mathrm{en}}(w)/F_2^{\mathrm{ep}}(w) \leqslant 4. \tag{1.64}$$

This is the Majumdar–Nachtmann inequality (Majumdar 1971, Nachtmann 1971). Experimentally, the situation is illustrated by Fig. 20. The data at $w \simeq 0{\cdot}8$ appear dangerously close to violating the lower bound. However, there is an indication that, beyond $w = 0{\cdot}8$, the points may be levelling off just above the value 0·25 for the ratio in question (Sogard 1972).

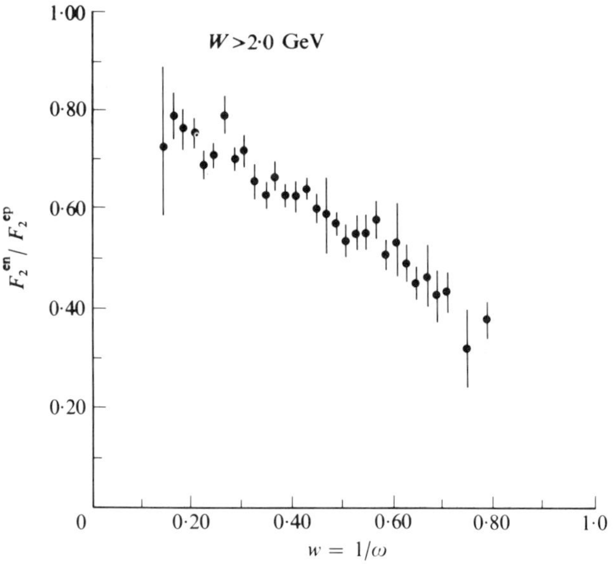


FIG. 20. $F_2^{\mathrm{en}}/F_2^{\mathrm{ep}}$ versus the scaling variable $w$ (Sogard 1972).

It is clear from above that the lower bound in eqn (1.64) is closely saturated. Then it becomes interesting to obtain another bound due to Nachtmann. First, note that

$$\frac{4F_2^{\mathrm{en}} - F_2^{\mathrm{ep}}}{F_2^{\nu\mathrm{p}}} = \frac{\frac{15}{9}(u_{\mathrm{q_n}}^{\mathrm{p}} + u_{\mathrm{q_{\bar{n}}}}^{\mathrm{p}}) + \frac{3}{9}(u_{\mathrm{q_\lambda}}^{\mathrm{p}} + u_{\mathrm{q_{\bar{\lambda}}}}^{\mathrm{p}})}{2(u_{\mathrm{q_n}}^{\mathrm{p}} + u_{\mathrm{q_{\bar{n}}}}^{\mathrm{p}})} = \frac{5(u_{\mathrm{q_n}}^{\mathrm{p}} + u_{\mathrm{q_{\bar{n}}}}^{\mathrm{p}}) + (u_{\mathrm{q_\lambda}}^{\mathrm{p}} + u_{\mathrm{q_{\bar{\lambda}}}}^{\mathrm{p}})}{6(u_{\mathrm{q_n}}^{\mathrm{p}} + u_{\mathrm{q_{\bar{n}}}}^{\mathrm{p}})}. \tag{1.65}$$

As explained earlier, in so far as the isospin content is concerned we can take $u_{\mathrm{q_n}}^{\mathrm{p}} \sim \langle \mathrm{p}|\mathrm{q_n^\dagger q_n}|\mathrm{p}\rangle$, etc. So let us construct a matrix $M_{\alpha r,\beta s} \sim \langle r|\mathrm{q}_\alpha^\dagger \mathrm{q}_\beta|s\rangle$ (where

$r, s$ can be p, n) and investigate the relations among its elements by considering the effect of the isospin-lowering operator $I_-$. Noting that $[I_-, \mathrm{q_n}] = -\mathrm{q_p}$ and $[I_-, \mathrm{q_p^\dagger}] = \mathrm{q_n^\dagger}$, we have

$$\begin{aligned}
\langle \mathrm{p}|\mathrm{q_p^\dagger q_n}|\mathrm{n}\rangle &= \langle \mathrm{p}|\mathrm{q_p^\dagger q_n} I_-|\mathrm{p}\rangle \\
&= \langle \mathrm{p}|\mathrm{q_p^\dagger q_p}|\mathrm{p}\rangle + \langle \mathrm{p}|\mathrm{q_p^\dagger} I_- \mathrm{q_n}|\mathrm{p}\rangle \\
&= u^{\mathrm{p}}_{\mathrm{q_p}} + \langle \mathrm{p}|I_- \mathrm{q_p^\dagger q_n}|\mathrm{p}\rangle - \langle \mathrm{p}|\mathrm{q_n^\dagger q_n}|\mathrm{p}\rangle \\
&= u^{\mathrm{p}}_{\mathrm{q_p}} - u^{\mathrm{p}}_{\mathrm{q_n}}.
\end{aligned}$$

Proceeding in this way, we easily obtain all the elements of $M$, and can write

$$M = \begin{bmatrix} u^{\mathrm{p}}_{\mathrm{q_p}} & 0 & 0 & u^{\mathrm{p}}_{\mathrm{q_p}} - u^{\mathrm{p}}_{\mathrm{q_n}} \\ 0 & u^{\mathrm{p}}_{\mathrm{q_n}} & 0 & 0 \\ 0 & 0 & u^{\mathrm{p}}_{\mathrm{q_n}} & 0 \\ u^{\mathrm{p}}_{\mathrm{q_p}} - u^{\mathrm{p}}_{\mathrm{q_n}} & 0 & 0 & u^{\mathrm{p}}_{\mathrm{q_p}} \end{bmatrix}. \tag{1.66}$$

This matrix has the eigenvalues $u^{\mathrm{p}}_{\mathrm{q_n}}$ (three-fold degenerate) and $2u^{\mathrm{p}}_{\mathrm{q_p}} - u^{\mathrm{p}}_{\mathrm{q_n}}$. Similarly, defining $\overline{M}_{\alpha r, \beta s} \sim \langle r|\bar{q}^\dagger_\alpha \bar{q}_\beta|s\rangle$, we obtain

$$\overline{M} = \begin{bmatrix} u^{\mathrm{p}}_{\mathrm{q_{\bar p}}} & 0 & 0 & 0 \\ 0 & u^{\mathrm{p}}_{\mathrm{q_{\bar n}}} & u^{\mathrm{p}}_{\mathrm{q_{\bar n}}} - u^{\mathrm{p}}_{\mathrm{q_{\bar p}}} & 0 \\ 0 & u^{\mathrm{p}}_{\mathrm{q_{\bar n}}} - u^{\mathrm{p}}_{\mathrm{q_{\bar p}}} & u^{\mathrm{p}}_{\mathrm{q_{\bar n}}} & 0 \\ 0 & 0 & 0 & n^{\mathrm{p}}_{\mathrm{q_{\bar p}}} \end{bmatrix}, \tag{1.67}$$

which has the eigenvalues $u^{\mathrm{p}}_{\mathrm{q_{\bar p}}}$ (three-fold degenerate) and $2u^{\mathrm{p}}_{\mathrm{q_{\bar n}}} - u^{\mathrm{p}}_{\mathrm{q_{\bar p}}}$. Now if $a^\dagger$ is the creation operator, $|s\rangle = a^\dagger_s|0\rangle$. Thus $M_{\alpha r, \beta s} \sim \langle 0|a_r q^\dagger_\alpha q_\beta a^\dagger_s|0\rangle$. Therefore, if we choose any set of vectors $C_{\beta s}$ spanning the SU(2) space concerned, we obtain

$$C^*_{r\alpha} M_{\alpha r, \beta s} C_{\beta s} \sim \langle 0| |cqa^\dagger|^2 |0\rangle \geqslant 0.$$

But this inequality can be satisfied only if $M$ is a positive matrix invariant under SU(2), and hence its eigenvalues have to be positive. Thus we have

$$2u^{\mathrm{p}}_{\mathrm{q_p}} - u^{\mathrm{p}}_{\mathrm{q_n}} \geqslant 0. \tag{1.68}$$

Similarly, from the positivity of $\overline{M}$, we obtain

$$2u^{\mathrm{p}}_{\mathrm{q_{\bar n}}} - u^{\mathrm{p}}_{\mathrm{q_{\bar p}}} \geqslant 0. \tag{1.69}$$

Taking advantage of eqn (1.69), we can write

$$u^{\mathrm{p}}_{\mathrm{q_n}} + u^{\mathrm{p}}_{\mathrm{q_{\bar p}}} \leqslant u^{\mathrm{p}}_{\mathrm{q_n}} + 2u^{\mathrm{p}}_{\mathrm{q_{\bar n}}} \leqslant 2(u^{\mathrm{p}}_{\mathrm{q_n}} + u^{\mathrm{p}}_{\mathrm{q_{\bar n}}}),$$

where we have used the positivity of the *u*s themselves. From eqn (1.65) and this last inequality we have

$$(4F_2^{\text{en}} - F_2^{\text{ep}})/F_2^{\nu\text{p}} \geqslant \tfrac{5}{12}. \tag{1.70a}$$

Eqn (1.70a) should be particularly interesting near $w \simeq 1$, where $4F_2^{\text{en}} - F_2^{\text{ep}}$ is known to be small. Eqns (1.68) and (1.69), coupled to eqns (1.21) and (1.22), also lead trivially to the inequality

$$F_2^{\bar{\nu}\text{p}}/F_2^{\nu\text{p}} \geqslant 2. \tag{1.70b}$$

Eqns (1.70a) and (1.70b) are known as Nachtmann inequalities (Nachtmann 1971, 1972*a*). Various consequences (of experimental interest) of the quark–parton model, along with the role played by certain questionable assumptions, are given in Table 1.1. We have deliberately concentrated on those results which follow from general symmetry arguments and the like. We should mention, however, that there are also more speculative dynamical models (with specific forms for the parton distribution functions) that have

TABLE 1.1
*Parton results and assumptions with GMZ quarks*

| Relation | Three core-quarks + uniform-sea assumption | Symmetric distribution assumption | Absence of gluons |
|---|---|---|---|
| Gottfried sum rule | Yes | No | No |
| Adler sum rule | No | No | No |
| Gross–Llewellyn-Smith sum rule | No | No | No |
| $\varepsilon_{\text{g}} = 1 + \int_0^1 \text{d}w \left\{\tfrac{3}{4}F_2^{\nu\text{p}+\nu\text{n}}(w) - \tfrac{9}{2}F_2^{\text{ep}+\text{en}}(w)\right\}$ | No | No | No |
| $\int_0^1 \text{d}w\, F_2^{\text{ep}}(w) = \dfrac{2}{9} + \dfrac{1}{3}\left\langle \dfrac{1}{l} \right\rangle_{\text{p}} \geqslant \dfrac{2}{9}$ | Yes | Yes | Yes |
| $\int_0^1 \text{d}w\, F_2^{\text{en}}(w) = \tfrac{2}{9}$ | Yes | Yes | Yes |
| $\int_0^1 \text{d}w\, F_2^{\text{ep}-\text{en}}(w) = \dfrac{1}{3}\left\langle \dfrac{1}{l} \right\rangle_{\text{p}} \geqslant 0$ | Yes | Yes | No |
| Llewellyn-Smith equality and inequality | No | No | No |
| Majumdar–Nachtmann inequality | No | No | No |
| Nachtmann inequalities | No | No | No |
| $\int_0^1 \dfrac{\text{d}w}{w}\left\{F_2^{\text{ep}}(w) - F_2^{\text{en}}(w)\right\} = \tfrac{1}{3} + \tfrac{2}{9}\langle l \rangle_{\text{p-n}}$ | Yes | No | Yes |

stronger predictions and attempt detailed fits with experimental curves (Kuti and Weisskopf 1971).

*The nature of quark–partons*

The results discussed so far are in the standard Gell-Mann and Zweig (GMZ) quark model. We can consider some of the relations derived earlier (e.g. CCSR and BCSR) in the Sakata model, having a p, n, $\lambda$ triplet with charges $\lambda_p = 1$, $\lambda_n = 0 = \lambda_\lambda$, but the results violate experiment (see Exercise 1.6). Another famous constituent model is the Han–Nambu (HN) three-triplet model (Kokkedee 1969), which we now describe briefly, mainly quoting the results. There are other competing quark models, but the Han–Nambu model is chosen here for the sake of definiteness. We shall compare the HN and the GMZ models in the context of deep inelastic $l$N scattering, although similar comparisons can also be made with respect to the others (Budny 1973, Choudhury 1973). First, the following points may be noted. (1) The use of Fermi statistics for quarks is successful in the construction of the observed hadron spectra in the HN model, whereas the standard GMZ model has difficulty in reconciling Fermi statistics with observed hadron spectroscopy, and consequently we have to adopt either Bose statistics or para-Fermi statistics of rank 3 (equivalently, three 'colours') for the quarks (Gell-Mann 1972). (2) The treatment of the $\pi^0 \to 2\gamma$ decay problem via the quark triangle diagram of the Adler anomaly also favours a triplet model such as that of HN, or alternatively a GMZ model with paraquarks of rank 3 (Bardeen, Fritzsch, and Gell-Mann 1972).

In the HN model there are three fundamental triplets of quarks and the same of antiquarks:

$$q_{pi}, q_{ni}, q_{\lambda i}; q_{\bar{p}i}, q_{\bar{n}i}, q_{\bar{\lambda}i} \qquad (i = 1, 2, 3).$$

These carry—along with the indices p, n, $\lambda$—the usual values for the quantum numbers $B$(baryon number) and $S$(strangeness). However, their charge quantum numbers are

$$\begin{aligned} Q_{q_{p1}} &= Q_{q_{p2}} = +1, \quad Q_{q_{p3}} = 0; \\ Q_{q_{n1}} &= Q_{q_{n2}} = 0, \quad\;\; Q_{q_{n3}} = -1; \\ Q_{q_{\lambda 1}} &= Q_{q_{\lambda 2}} = 0, \quad\;\; Q_{q_{\lambda 3}} = -1; \end{aligned} \tag{1.71}$$

with $Q_{\bar{q}} = -Q_q$. Further, they carry a new quantum number $C$, called charm,

$$C = \begin{cases} +1 & \text{for } i = 1, \\ -2 & \text{for } i = 2, \\ +1 & \text{for } i = 3. \end{cases} \qquad C_{\bar{q}} = -C_q. \tag{1.72}$$

Usual hadrons have $C = 0$. Charmed particles (which will be strongly stable if charm is conserved in strong interactions) are not yet seen, presumably because they are of very high mass. Thus, in the HN model,

$$|\pi^+\rangle = \frac{1}{\sqrt{3}}|q_{p1}q_{\bar{n}1} + q_{p2}q_{\bar{n}2} + q_{p3}q_{\bar{n}3}\rangle \qquad (= |q_p q_{\bar{n}}\rangle \text{ in GMZ}),$$

$$|p\rangle = \frac{1}{\sqrt{6}}|q_{p1}q_{p2}q_{n3} + q_{p2}q_{p3}q_{n1} + q_{p3}q_{p1}q_{n2}$$
$$-q_{p1}q_{n3}q_{p2} - q_{p2}q_{n1}q_{p3} - q_{p3}q_{n2}q_{p1}\rangle$$
$$(= |q_p q_p q_n\rangle \text{ in GMZ}).$$

We now come to two important points:

*Higher symmetry and consequent modification of the sum rules.* In this scheme we have the usual SU(3) symmetry operating along the $p \to n \to \lambda$ chain. But, in addition, there is now an extra symmetry group $\widetilde{SU}(3)$ operating on the indices $i(= 1, 2, 3)$. So the model has the higher symmetry $SU(3) \times \widetilde{SU}(3)$ and consequently not only the ordinary isospin SU(2) but the larger $SU(2) \times \widetilde{SU}(2)$. Instead of isospin invariance, the full invariance under the latter has to be used. This leads to the modification of a number of results derived earlier, in case the quarks are of the Han–Nambu type. Without going into the details here, we merely quote the results (Budny 1972). The Adler sum rule

$$\int_0^1 \frac{dw}{w}\{F_2^{\bar{\nu}p}(w) - F_2^{\nu p}(w)\} = 2 \tag{1.73}$$

is unchanged. The Gross–Llewellyn-Smith sum rule becomes

$$\int_0^1 dw\,\{F_3^{\bar{\nu}p}(w) + F_3^{\nu p}(w)\} = -10, \tag{1.74}$$

where the right-hand side is $-6$ with GMZ quarks. The Llewellyn-Smith isospin equality and inequality (eqns (1.35)) are also changed; e.g. the latter becomes

$$F_{1,2}^{ep} + F_{1,2}^{en} \geqslant \tfrac{1}{2}(F_{1,2}^{\nu p} + F_{1,2}^{\nu n}). \tag{1.75}$$

The Majumdar–Nachtmann inequality (eqn (1.64)) now becomes

$$0 \leqslant F_2^{en}(w)/F_2^{ep}(w) \leqslant \infty. \tag{1.76}$$

If we use symmetric momentum distribution, then this changes to

$$\tfrac{1}{2} \leqslant F_2^{\text{en}}(w)/F_2^{\text{ep}}(w) \leqslant 2, \tag{1.77}$$

in violent disagreement with experiment. However, perhaps this merely means that we may dispense with the symmetric momentum distribution assumption.

*Lipkin's argument* (Lipkin 1972). It may be asserted on the basis of the following argument that, for the production of regular hadrons induced by a single current at present energies, we can never distinguish between the GMZ and HN models. In the HN model each hadron has a multi-component wavefunction, and the individual charge of quarks or antiquarks varies from part to part. Thus for the $\pi^+$, say, the instantaneous quark or antiquark charges are integral, but the average quark charge is $\frac{2}{3}$ and the average antiquark charge is $\frac{1}{3}$, as in the GMZ model. The same is true also for the proton. This implies the existence of charge-exchange interactions (among the quarks and antiquarks) causing fluctuations of the parton charge, leading to $\frac{1}{3}$-integral charges on the average. Hence, unless the time scale of our measurement is somewhat smaller than typical fluctuation times, we cannot distinguish the HN model from the GMZ model. But what does it mean to have measurements with this kind of time scale? We have to phrase the argument in experimentally more meaningful terms. Again, take the example of the $\pi^+$. With three $\text{q}_\text{p}$s and three $\text{q}_\text{n}$s we can construct nine zero-isospin states, of which only $\pi^+$ is the uncharmed singlet, the other eight being charmed particles forming an octet under $\widetilde{\text{SU}}(3)$. A similar statement holds for all ordinary observed hadrons, which are $\widetilde{\text{SU}}(3)$ singlets with their charmed counterparts in the corresponding $\widetilde{\text{SU}}(3)$ multiplets. This crucial fact (that ordinary hadrons are $\widetilde{\text{SU}}(3)$ singlets) is all that is needed to proceed further, rather than requiring any explicit form for the wavefunctions. Thus an additional sea of $\bar{\text{q}}\text{q}$ pairs, so long as it is an $\widetilde{\text{SU}}(3)$ singlet, does not affect the following argument.

Take the charge operator. We note that $Q^{\text{HN}}$ can be broken up into two parts:

$$Q^{\text{HN}} = Q^{\text{GMZ}} + Q^{\text{A}},$$

where $Q^{\text{GMZ}}_{\text{q}_\text{p}} = \frac{2}{3}$, $Q^{\text{GMZ}}_{\text{q}_\text{n}} = Q^{\text{GMZ}}_{\text{q}_\lambda} = -\frac{1}{3}$ (independent of 1, 2, 3), and $Q^{\text{A}}$ is the difference between the instantaneous charge and the average parton charge $= Q^{\text{HN}} - Q^{\text{GMZ}}$. Hence

$$Q_1^{\text{A}} = Q_2^{\text{A}} = \tfrac{1}{3}, \qquad Q_3^{\text{A}} = -\tfrac{2}{3} \qquad \text{(independent of p, n, } \lambda). \tag{1.78}$$

Therefore $Q^{\text{GMZ}}$ is an operator belonging to an octet in SU(3) and to a singlet in $\widetilde{\text{SU}}(3)$, whereas $Q^{\text{A}}$ belongs to a singlet in SU(3) and to an octet in $\widetilde{\text{SU}}(3)$.

Now let $\alpha$ refer to (p, n, $\lambda$) and $i$ to (1, 2, 3). We can construct the hadron electromagnetic current as follows:

$$J_\mu^{\text{EM}} \sim \sum_{i,\alpha} \bar{q}_{\alpha i}\gamma_\mu Q_{\alpha i}^{\text{HN}} q_{\alpha i}$$

$$= \sum_{\alpha,i} \bar{q}_{\alpha i}\gamma_\mu Q_\alpha^{\text{GMZ}} q_{\alpha i} + \sum_{\alpha,i} \bar{q}_{\alpha i}\gamma_\mu Q_i^{\text{A}} q_{\alpha i}.$$

Thus

$$J_\mu^{\text{EM}} = J_\mu^{\text{GMZ}} + J_\mu^{\text{A}}, \tag{1.79}$$

where $J_\mu^{\text{GMZ}}$ and $J_\mu^{\text{A}}$ have the same properties under SU(3) and $\widetilde{\text{SU}}(3)$ as $Q^{\text{GMZ}}$ and $Q^{\text{A}}$ respectively. For deep inelastic electron–hadron scattering, we need to calculate sums over matrix elements squared, such as

$$\sum_n C_n \langle B|J_\mu^{\text{EM}}|n\rangle\langle n|J^{\mu,\text{EM}}|B\rangle = \sigma, \tag{1.80}$$

say, where $B$ is a usual hadron. Since the latter is in an $\widetilde{\text{SU}}(3)$ singlet, only transitions to uncharmed states $|n_0\rangle$ (singlet in $\widetilde{\text{SU}}(3)$) can come from the current $J_\mu^{\text{GMZ}}$, whereas only transitions to charmed states $|n_8\rangle$ (octet in $\widetilde{\text{SU}}(3)$) can come from the $J_\mu^{\text{A}}$ part. Therefore we may write $\sigma = \sigma_0 + \sigma_8$, where

$$\begin{aligned} \sigma_0 &= \sum_{n_0} C_{n_0}\langle B|J_\mu^{\text{GMZ}}|n_0\rangle\langle n_0|J^{\mu,\text{GMZ}}|B\rangle, \\ \sigma_8 &= \sum_{n} C_{n_8}\langle B|J_\mu^{\text{A}}|n_8\rangle\langle n_8|J^{\mu,\text{A}}|B\rangle. \end{aligned} \tag{1.81}$$

Hence, if we are producing only uncharmed particles, as at present energies, $\sigma_8 = 0$. Thus even if the electromagnetic current $J_\mu^{\text{EM}}$ has two components $J_\mu^{\text{GMZ}}$ and $J_\mu^{\text{A}}$, the partons 'see' only the $J_\mu^{\text{GMZ}}$ part. In other words, if no charmed particles are produced, the time of interaction is too large to see the above-mentioned charge fluctuation. On the other hand, if charmed particles are produced, they can be detected directly, to support the HN model without recourse to sum rules.

The upshot of the above discussion is as depicted in Fig. 21—namely, that the production of charmed particles will herald a new scaling domain where eqns (1.73)–(1.76) hold in the HN model. Quite generally—in the region of new high-mass particles—we expect a fundamental length proportional to the inverse of the mass to enter and violate scaling. The idea of a breakdown in scaling in some high-energy region is actually a recurrent feature in many models, where special high-mass channels among the deep inelastic hadronic final states are supposed to play a fundamental role.

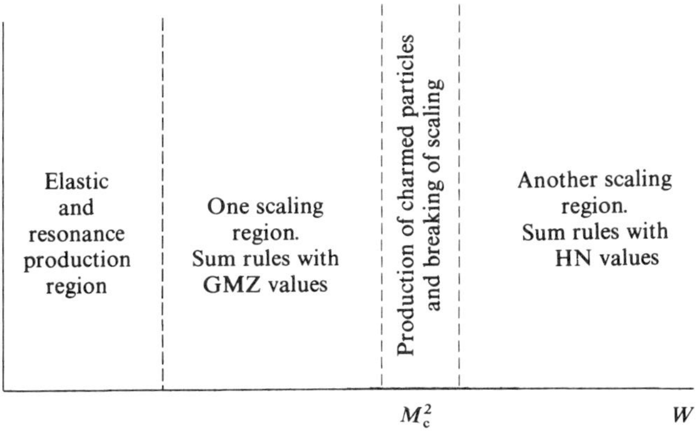


FIG. 21. Transition from GMZ to HN sum rules due to charmed particles of mass $M_c$.

## Exercises

1.1. Show by helicity arguments in the Breit frame (Fig. 16) that there is no coupling between longitudinal photons and spin-$\frac{1}{2}$ partons, as well as none between transverse photons and spin-0 partons in the limit of large momentum transfer.

1.2. Derive eqn (1.18) in the approximation that the Cabibbo angle $\theta_C$ is zero.

1.3. Obtain eqn (1.22), following the derivation of eqn (1.21).

1.4. If partons have spin zero, show that $F_1^{\nu\bar{\nu}} = 0$.

1.5. Derive the Llewellyn-Smith inequality (eqn (1.35b)). Under what conditions does it reduce to an equality?

1.6. How do the relations CCSR and BCSR, as well as the equality and inequality of Llewellyn-Smith, look like in the triplet model of Sakata ($\lambda_p = 1$, $\lambda_n = 0$, $\lambda_\lambda = 0$)? Show that the said model is incompatible with the data on $\int_0^1 \mathrm{d}w\, F_2^{\mathrm{ep+en}}(w)$ and $\int_0^1 \mathrm{d}w\, F_2^{\nu\mathrm{p}+\nu\mathrm{n}}(w)$.

1.7. Show that the ratio $\int_0^1 \mathrm{d}w\, w\{2F_1^{\nu\mathrm{p}+\nu\mathrm{p}}(w) - F_3^{\nu\mathrm{p}+\nu\mathrm{p}}(w)\}/\int_0^1 \mathrm{d}w\{2F_1^{\nu\mathrm{p}+\nu\mathrm{p}}(w) + F_3^{\nu\mathrm{p}+\nu\mathrm{p}}(w)\}w$ lies between $\frac{1}{3}$ and 3 if the partons are quarks and antiquarks of GMZ and possible gluons.

# 2

# THE THEORETICAL BASIS OF THE PARTON PICTURE

## A critique of kindergarten parton calculations

THE validity of certain assumptions, taken for granted throughout the kindergarten parton calculations, will now be critically discussed. In this chapter we shall confine ourselves to the case of deep inelastic electron–proton scattering; however, our arguments admit of ready generalization to include the corresponding reactions induced by neutrinos and antineutrinos. Consider first the following objections to and criticisms of the said calculations. (1) We have always assumed that all the partons move forward in time, i.e. that the longitudinal fraction $x_i$ (where $\sum_i x_i = 1$) are in the range $0 < x_i < 1$. However, there is no *a priori* reason why some of the $x_i$s cannot be less than zero and others accordingly greater than unity. (2) We have taken a point vertex for the $\gamma^*$–parton interaction. However, in general, we could have other diagrams, as shown in Fig. 22. (3) The whole question of final-state interactions has been glossed over. The computation of $W^{e}_{\mu\nu}$, for example, has been done assuming the equality as depicted in Fig. 12.

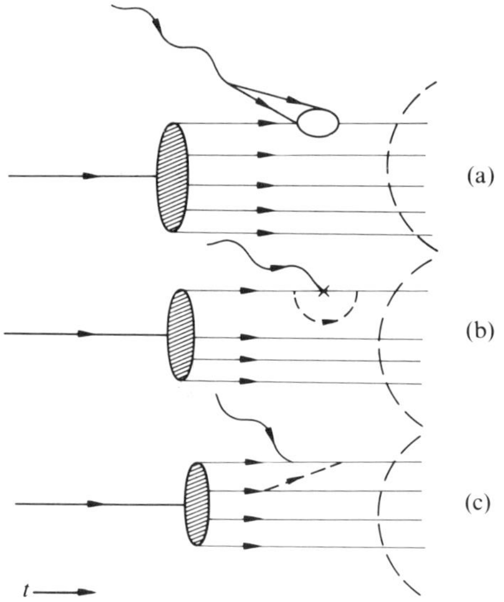


FIG. 22. Illustrative supplementary parton diagrams.

In this figure, $\sum$ refers to the incoherent summation wherein the partons somehow take care of themselves after the photon–parton interaction. (4) In carrying out the kindergarten calculations we have not explained the non-production of actual partons. Bjorken scaling sets in at experimentally $\sqrt{Q^2_{\min}} \simeq 1$ GeV, and clearly the parton mass has to be smaller than this. Then why are they not knocked out, and their interesting properties seen directly? Indeed, what is the connection between the constituent partons and the known hadrons? (5) Finally, we have provided no theoretical understanding of the cut-off in the transverse momenta of the partons.

In this chapter we want to consider some of the above questions. We cannot guarantee satisfactory answers to all of them, but can perhaps provide the reader with sufficient material for further meditation. First, we need a deeper and sounder theoretical framework. So far attempts towards constructing such a framework have been only partially successful. They have been based on the two main theoretical approaches in particle physics: (1) Lagrangian field theory and (2) $S$-matrix theory. Thus, for example, a parton model for lepton–hadron processes has been developed by Drell, Levy, and Yan (DLY) (Drell and Yan 1971) using field theory and a model based on $S$-matrix theory was advanced by Landshoff, Polkinghorne, and Short (LPS) (Landshoff and Polkinghorne 1972). Broadly similar basic results with interesting differences in detail have been achieved by these models. The field-theoretical model has been developed on the basis of old-fashioned perturbation theory in the infinite momentum frame with interactions of the form $g\bar{\psi}\psi\phi$ or $g\bar{\psi}\gamma_5\psi\phi$ with bare hadrons such as nucleons, $\sigma$s, and $\pi$s; it is, in principle, also extendable to quarks and gluons. In the other case, the photon is decomposed via a point-like vertex into a parton–antiparton pair, and the parton–hadron scattering amplitude is treated using techniques analogous to Gribov's Reggeon calculus (Gribov 1968). One qualification, however, will be in order at this point. The DLY scheme cannot really qualify for a genuine perturbative Lagrangian field theory because the virtual transverse momenta in all loop diagrams are cut off arbitrarily. Similarly, the LPS model is not an authentic $S$-matrix theoretical model, since its necessary assumption of a point-like dissociation of the photon into a parton–antiparton pair suggests an interaction Lagrangian. A comparison of the two approaches is given below.

| DLY | LPS |
|---|---|
| (1) Partially covariant (valid in a class of $P_z \to \infty$ frames). | (1) Completely covariant. |
| (2) Transverse momentum cut-off on the partons is necessary and is an *ad hoc* input. | (2) No such cut-off is necessary; however, softness of the parton–hadron scattering amplitude is taken as a dictum (i.e. the parton–hadron |

| | |
|---|---|
| | amplitude is assumed to go to zero sufficiently rapidly as the parton mass becomes large). Furthermore, the parton propagator is assumed to behave as a free-particle propagator at large mass. |
| (3) The point-like property of the $\gamma^*$–parton vertex is an output. | (3) The assumption of a point-like $\gamma^*$–parton vertex is an input. |
| (4) Perturbative approach to strong interactions present. | (4) Non-perturbative approach to the strong-interaction part. |
| (5) Close in spirit to Feynman's original picture and 'derives' impulse approximation and hence scaling. | (5) Relation to Feynman's picture is rather obscure, but scaling is obtained. |
| (6) Gives a pseudo-explanation of the non-appearance of real partons among the final states. | (6) Cannot explain why actual partons are not produced. |
| (7) Relation to duality and Regge ideas is obscure. | (7) Can connect with Regge theory and duality by incorporating Regge trajectories in a realistic way. |

Because the DLY model is closer to Feynman's intuitive arguments, it will be discussed in some detail. We shall, however, present a brief review of the LPS approach also.

**The model of Drell, Levy, and Yan**

Here the total Hamiltonian is split up into two parts—the free solvable part and the interaction

$$H = H_0 + H_{\rm I}. \tag{2.1}$$

The renormalized hadronic electromagnetic current $J_\mu^{\rm EM}(x)$ can be 'undressed' by going into the interaction picture, where

$$J_\mu^{\rm EM}(x) = U^\dagger(t, -\infty) j_\mu^{\rm EM}(x) U(t, -\infty). \tag{2.2}$$

In eqn (2.2) $j_\mu^{\rm EM}(x)$ is the bare electromagnetic current and can be written in terms of the field operators as

$$j_\mu^{\rm EM}(x) = \begin{cases} e\bar{\psi}\gamma_\mu\psi \text{ (fermions)}, \\ e\varphi^\dagger \overleftrightarrow{\partial}_\mu \varphi \text{ (bosons)}, \end{cases} \tag{2.3}$$

where $e$ is the renormalized charge, and $\psi$, $\varphi$ are free-field operators. The operator $U(t, t')$ is of the following form (Bjorken and Drell 1965):

$$U(t, t') = P \exp\left\{-\mathrm{i} \int_{t'}^{t} H_{\mathrm{I}}(\tau)\, \mathrm{d}\tau\right\}. \tag{2.4}$$

Note that the $S$-matrix is

$$S = \lim_{\substack{t \to \infty \\ t' \to -\infty}} U(t, t'), \tag{2.5}$$

where the limit is taken in a weak sense.

We can now write, from eqn (I.3),

$$\begin{aligned} W^{\mathrm{e}}_{\mu\nu} &= (2\pi)^2 \frac{p_0}{M} \sum_n \langle p|J^{\mathrm{EM}}_{\mu}(0)|n\rangle \langle n|J^{\mathrm{EM}}_{\nu}(0)|p\rangle (2\pi)^4 \delta^{(4)}(p+q-p_n) \\ &= (2\pi)^2 \frac{p_0}{M} \sum_n \langle Up|j^{\mathrm{EM}}_{\mu}(0)|U_n\rangle \langle U_n|j^{\mathrm{EM}}_{\nu}(0)|Up\rangle \times \\ &\quad \times (2\pi)^4 \delta^{(4)}(p+q-p_n), \end{aligned} \tag{2.6}$$

where

$$|Up\rangle = U(0, -\infty)|p\rangle, \tag{2.7}$$

and so on. Since $j^{\mathrm{EM}}_{\mu}(0)$ is a one-body operator associated with $H_0$, it can only connect—via a point-like vertex—some one-particle state (say with momentum $p$) projected from $|Up\rangle$ with a similar one-particle state projected out from $|Un\rangle$. Now, by expanding the right-hand side of eqn (2.4), we can decompose $|Up\rangle$ into a superposition of many particle states as follows:

$$\begin{aligned} |Up\rangle &= U(0, -\infty)|p\rangle \\ &= \sum_{l=0}^{\infty} \frac{(-\mathrm{i})^l}{l!} \int_{-\infty}^{0} \mathrm{d}\tau_1 \ldots \int_{-\infty}^{0} \mathrm{d}\tau_l\, T H_{\mathrm{I}}(\tau_1) \ldots H_{\mathrm{I}}(\tau_l)|p\rangle. \end{aligned} \tag{2.8}$$

This operation is called the undressing of the proton. A term-by-term calculation with appropriate insertion of complete sets of states, which are eigenstates of $H_0$ (these correspond to the parton states $\{|l\rangle\}$ of Chapter 1), leads to (see Exercise 2.1)

$$\begin{aligned} |Up\rangle = \sqrt{Z_2}\Bigg\{|p\rangle &+ \sum_m{}' \frac{|m\rangle\langle m|H_{\mathrm{I}}|p\rangle}{p^0 - p^0_m} + \sum_{m,r}{}' \frac{|m\rangle\langle m|H_{\mathrm{I}}|r\rangle\langle r|H_{\mathrm{I}}|p\rangle}{(p^0 - p^0_m)(p^0 - p^0_r)} + \\ &+ \ldots\Bigg\}, \end{aligned} \tag{2.9}$$

where $p^0$, $p_m^0$ are the energy eigenvalues of the free particle states, $\sum'$ means summation over all states except $|p\rangle$, and $Z_2$ is the wavefunction renormalization constant of the proton. We can see in every matrix element that energy is not conserved, i.e. $p^0 \neq p_m^0$, although momentum is, i.e. $\mathbf{p} = \mathbf{p}_m = \ldots$. $Z_2$ is of the following form:

$$Z_2 = 1+(-\mathrm{i})^2 \int_{-\infty}^{0} \mathrm{d}\tau \int_{-\infty}^{\tau} \mathrm{d}\tau' \, \langle p|H_{\mathrm{I}}(\tau)H_{\mathrm{I}}(\tau')|p\rangle + \ldots$$
$$= 1+O(g^2), \tag{2.10}$$

where $g$ is the coupling constant associated with $H_{\mathrm{I}}$. We can check (see Exercise 2.2) term by term that the unitarity condition for $U$ is being satisfied, i.e.

$$\langle Up'|Up\rangle = \langle p|U^{\dagger}(0,-\infty)U(0,-\infty)|p\rangle = \langle p'|p\rangle = \delta^{(3)}(\mathbf{p}'-\mathbf{p}). \tag{2.11}$$

Let us study eqn (2.9) with an example from the pseudoscalar-meson–nucleon theory defined by the interaction $\mathscr{L}_{\mathrm{I}} = g\bar{\psi}\gamma_5\psi\varphi$. A typical undressing vertex is shown in Fig. 23. We have to discuss it by using old-fashioned

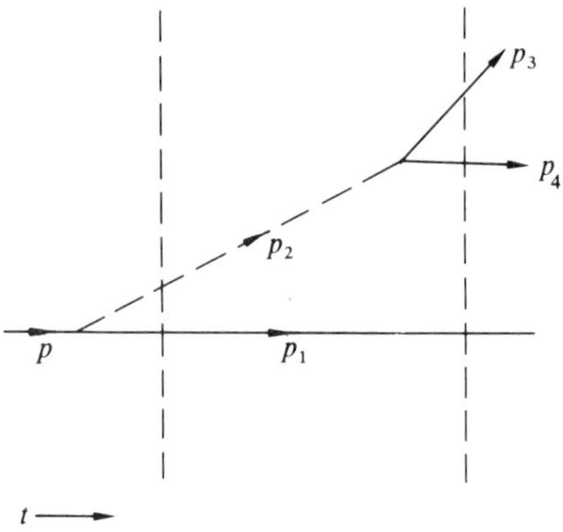


FIG. 23. Typical undressing vertex.

perturbation theory (OFPT), in the spirit of the argument given at the beginning of Chapter 1. The transition amplitude corresponding to this figure in OFPT is proportional to

$$\sqrt{\left(\frac{M}{p_0}\right)} \int \frac{\mathrm{d}^3 p_2}{2p_2^0} \delta^{(3)}(\mathbf{p}_3+\mathbf{p}_4-\mathbf{p}_2) g^2 \frac{\bar{u}(p_3)\gamma_5 v(p_4)}{p^0-p_1^0-p_3^0-p_4^0} \frac{\bar{u}(p_1)\gamma_5 u(p)}{p^0-p_1^0-p_2^0}. \tag{2.12}$$

In the expression given in eqn (2.12) the terms $\bar{u}(p_3)\gamma_5 v(p_4)/(p^0-p_1^0-p_3^0-p_4^0)$ and $\bar{u}(p_1)\gamma_5 u(p)/(p^0-p_1^0-p_2^0)$ are related to $\langle m|H_{\mathrm{I}}|r\rangle/(p^0-p_m^0)$ and $\langle r|H_{\mathrm{I}}|p\rangle/(p^0-p_r^0)$ respectively.

We now notice that in the case of elastic electron–proton scattering, the relevant matrix element is

$$\begin{aligned}\langle p'|J_\mu^{\mathrm{EM}}(0)|p\rangle &= \langle Up'|j_\mu^{\mathrm{EM}}(0)|Up\rangle \\ &= Z_2\{\langle p'|j_\mu^{\mathrm{EM}}(0)|p\rangle + O(g^2)\}. \end{aligned} \tag{2.13}$$

The point-like matrix element of the bare current does not decrease significantly with $Q^2$. Hence if we do want to have a fast fall-off of the elastic form factor at large $Q^2$, as observed (see the Introduction, p. 10), we must have the consistency requirement $Z_2 = 0$. Moreover, the terms of order $g^2$ and above do not have this point-like structure and will not cancel this first term. The vanishing of $Z_2$ means that the diagram shown in Fig. 24 cannot

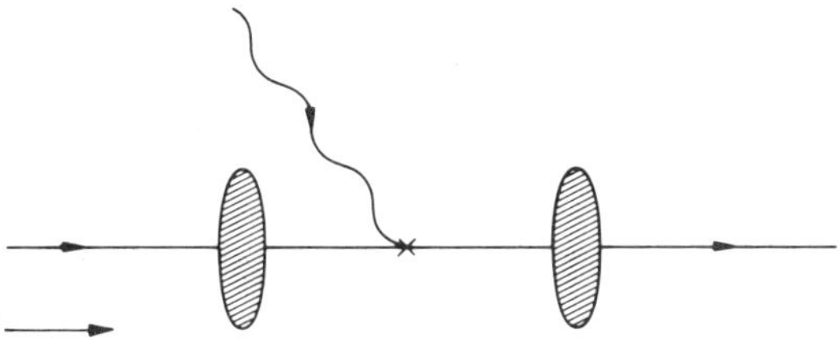


FIG. 24. Vanishing $Z_2$ diagram.

contribute. However, it does not mean—as we might naively conclude from eqn (2.9)—that the state $|Up\rangle$ is identically zero, leading to the vanishing of the inelastic structure functions. Had $|Up\rangle$ vanished identically, the elastic form factor would have been zero for all $q^2$; this is clearly not so. $F_{1,2}^{\mathrm{ep}}(0)$ are the coefficients of the charge and the anomalous magnetic moment and hence are intrinsic properties of the proton; these cannot vanish. Thus, in eqn (2.9) the higher-order terms must sum in such a way that the curly-bracketed quantity is of the order of $1/\sqrt{Z_2}$, so that although $Z_2$ is zero, $|Up\rangle$ is non-zero. This is a reasonable conjecture, but it is difficult to verify in any field-theoretical model because there are an infinite number of terms which need to be summed.

There are three important features of the DLY model worthy of comment. First, we obtain a proper understanding of the role of the infinite momentum frame. The second nice feature of the model is a theoretical justification for the assumption that the longitudinal fraction of $P$ carried by a parton is between 0 and 1 ($0 < x_i < 1$—see Chapter 1) which leads to the crucial results $E_p - E_{Up} \to 0$, $E_n - E_{Un} \to 0$, in analogy with eqns (1.2) and (1.3), thereby ensuring that eqn (2.6) may be rewritten as

$$\lim W_{\mu\nu}^{\mathrm{e}} = (2\pi)^2 \frac{p_0}{M} \int \mathrm{d}^4x\, \mathrm{e}^{\mathrm{i}q.x} \langle Up|j_\mu^{\mathrm{EM}}(x) j_\nu^{\mathrm{EM}}(0)|Up\rangle \tag{2.14}$$

(cf. eqns (1.4), (1.19), and (1.20)). Finally, a definite attempt is made to grapple with the question of final-state interactions among the partons after the $\gamma^*$–parton collision. We now look in detail at these points.

*Infinite momentum frames*

The current method of formulating the parton model chooses a 'true' infinite momentum frame (Drell and Yan 1971). In such a frame the momentum variable $P$ does not control any observable and goes to infinity much faster than any such quantity. In any event $M\nu$ and $Q^2$ must have leading independence from $P$ in the said limit. The nucleon four-momentum can be taken to be

$$p^\mu = \left(P + \frac{M^2}{2P}, 0, 0, P\right), \tag{2.15}$$

where

$$P = |p_z| \to \infty.$$

Writing $q^\mu = (q^0, \mathbf{q}_\perp, q^3)$, we have

$$M\nu = (q^0 - q^3)P + \frac{M^2}{2P} q^0, \tag{2.16a}$$

$$Q^2 = -(q^3 + q^0)(q^3 - q^0) + \mathbf{q}_\perp^2. \tag{2.16b}$$

In the DLY scheme, in order to ignore the photon-dissociation graphs Fig. 22(a), it is necessary that $q^0$ and $q^3$ vanish in the infinite momentum frame. For example, in the pion–nucleon theory there can be four lowest-order time-ordered graphs for electro-pion-production, as shown in Fig. 22. The graphs of Fig. 25(c) and (d) are of the photon-dissociation type and should not contribute. They are outside the spirit of the parton model because they involve the propagation characteristics of the virtual photon rather than an instantaneous photon–parton interaction at infinite momentum. The proof that (c) and (d) do not contribute if $q^0$ and $q^3$ go to zero as $P$ approaches infinity is somewhat detailed and will not be given here (Drell and Yan 1971). The point, however, is to have such constraints on $q^{0,3}$. A general form for $q$, therefore, is

$$q^\mu = \left(\frac{\alpha}{P}, \mathbf{q}_\perp, \frac{\beta}{P}\right), \tag{2.17}$$

with $\alpha - \beta = M\nu$ and $q_\perp^2 = Q^2 + O(1/P^2)$.

There are two popular choices:

$$\alpha = \tfrac{1}{4}(2M\nu - Q^2), \qquad \beta = -\tfrac{1}{4}(2M\nu + Q^2) \tag{2.18a}$$

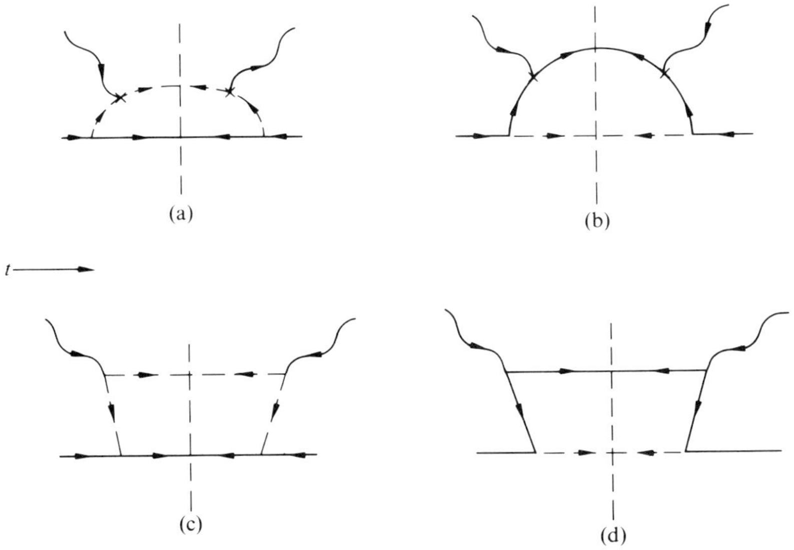


FIG. 25. Lowest-order time-ordered graphs for electro-pion-production.

or

$$\alpha = M\nu, \qquad \beta = 0. \tag{2.18b}$$

However, any choice of $\alpha$ and $\beta$ independent of $P$ and with $\alpha-\beta = M\nu$ is acceptable, thus generating a whole class of permitted frames. Notice that two observable quantities, the initial and final electron four-momenta, are of the following form:

$$p_e^\mu = \left(uP+\frac{2M_e^2+p_{e\perp}^2}{4uP}, \mathbf{p}_{e\perp}, uP-\frac{p_{e\perp}^2}{4uP}\right), \tag{2.19a}$$

$$p_e^{\prime\mu} = \left(uP+\frac{2M_e^2+p_{e\perp}^2+4\alpha u}{4uP}, \mathbf{p}_{e\perp}+\mathbf{q}_\perp, uP+\frac{4\beta u-p_{e\perp}^2}{4uP}\right), \tag{2.19b}$$

where $\mathbf{p}_{e\perp}\cdot\mathbf{q}_\perp = 0$, so that the condition $p_e^{\prime 2} = m_e^2 = p_e^2$ is maintained. In eqns (2.19) $p_{e\perp}^2$ and $u$ are related to $(p_e+p)^2$ by

$$s \equiv (p_e+p)^2 = (1+u)\left(M^2+\frac{m_e^2}{u}\right)+\frac{p_{e\perp}^2}{u}.$$

Thus the type of infinite momentum frame where the parton-analysis can be performed is clear from the above discussion.

*Forward-moving partons*

In the DLY approach the question of the validity of the assumption $0 < x_i < 1$ is connected with the role of $Z$-graphs (Sakurai 1967) in the infinite momentum frame. It is well known that a covariant $s$- or $u$-channel exchange Feynman diagram in the language of OFPT is the sum of a corresponding time-ordered pole diagram along with a $Z$-graph. For the $s$-channel Born diagram of Compton scattering the decomposition is shown in Fig. 26(a). The intermediate particle in the $Z$-graph is an antifermion with momentum $\mathbf{p}-\mathbf{k}$, going forward in time corresponding to the spinor $v(-\mathbf{p}-\mathbf{k})$. Similarly for the $u$-channel we have the equivalence of Fig. 26(b). In the

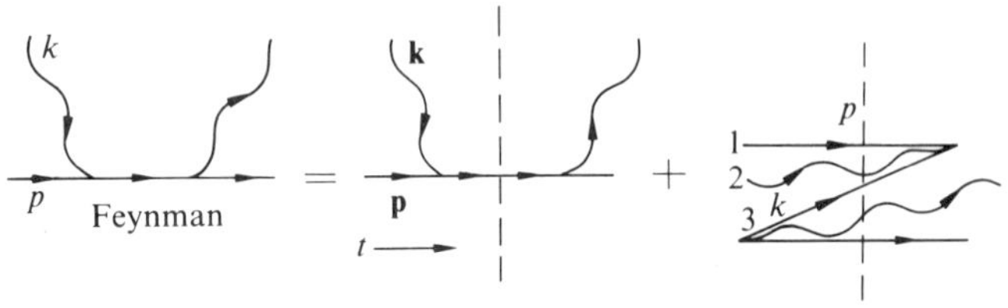


(a)

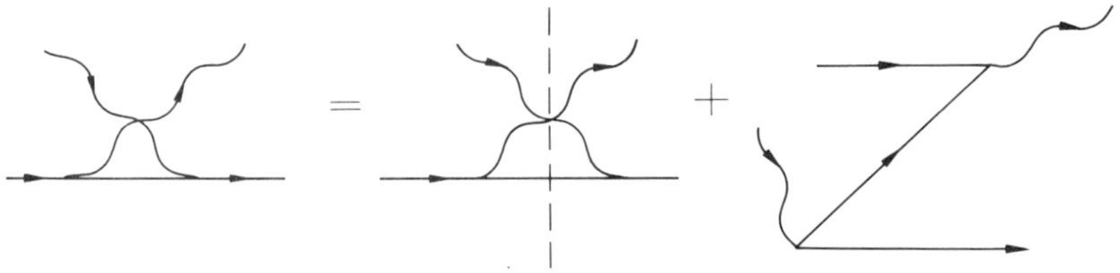

(b)

FIG. 26. (a) Time-ordered pole diagram/$Z$-graph decomposition of the $s$-channel Born diagram for Compton scattering. (b) As Fig. 26(a), but for the $u$-channel.

$Z$-graph of Fig. 26(a), say, 1, 2, and 3 cannot all appear to be travelling in the positive $z$-direction even in a longitudinally boosted frame. If 1 is going along the positive $z$-direction and 2 mainly in the transverse direction, 3 will go predominantly in the negative $z$-direction. This is unpleasant for the parton picture since the corresponding longitudinal fraction is becoming negative, i.e. $x_3$ is no longer between 0 and 1. Let us consider the problem in more detail. For spin-0 fields it has been shown (Weinberg 1966) that in an infinite momentum frame $Z$-graphs do not contribute. This is because, whereas each non-$Z$ (i.e. forward-moving) intermediate state is associated with an energy denominator of the form $1/O(M^2/P)$, the corresponding form for a $Z$-type (i.e. backward-moving) intermediate state is $1/O(P)$. The relative damping of $1/P^2$ for each $Z$-type energy denominator compared to a non-$Z$ one is sufficient to dispose of $Z$-graphs. We can demonstrate this in a $\varphi^3$ theory as follows. Let $m$ be the mass of the scalar particle. Consider the two

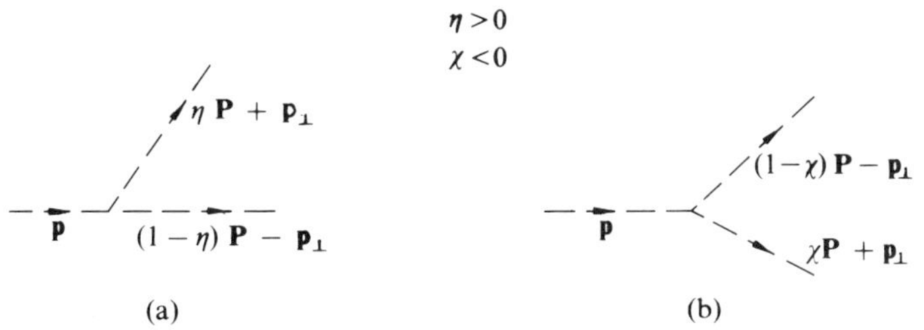


FIG. 27. (a) Non-Z-graph vertex. (b) Z-graph vertex.

vertices of Fig. 27. Fig. 27(a) corresponds to a non-$Z$ situation, and the corresponding energy denominator is

$$
\begin{aligned}
(p^0_{Up}-p^0)^{-1} &\simeq \left\{\eta P+\frac{k_\perp^2+m^2}{2\zeta P}+(1-\eta)P+\frac{k_\perp^2+m^2}{2(1-\eta)P}-P-\frac{m^2}{2P}\right\}^{-1}\\
&=\left\{\frac{k_\perp^2+m^2}{2P}\left(\frac{1}{\eta}+\frac{1}{1-\eta}\right)-\frac{m^2}{2P}\right\}^{-1}\\
&\simeq \frac{1}{O(m^2/P)}.
\end{aligned}
$$

On the other hand, for the $Z$-graph vertex of Fig. 27(b), we have the energy denominator

$$
\begin{aligned}
(p^0_{Up}-p^0)^{-1} &\simeq \left\{|\chi|P+\frac{k_\perp^2+m^2}{2|x|P}+|1-\chi|P+\frac{k_\perp^2+m^2}{2|1-\chi|P}-P-\frac{m^2}{2P}\right\}^{-1}\\
&=\left\{(1+2|\chi|)P+\frac{k_\perp^2+m^2}{2P}\left(\frac{1}{|\chi|}+\frac{1}{|1-\chi|}\right)-\frac{m^2}{2P}\right\}^{-1}\\
&\simeq \frac{1}{O(P)}.
\end{aligned}
$$

Unfortunately, for spin-$\frac{1}{2}$ fields things are not so simple. Now the numerators involve Dirac spinors and gamma matrices. In the infinite momentum limit these may give rise to certain compensating factors for the $Z$-graphs. If $p_z, p'_z \to \infty$ in the limit when $P \to \infty$, the bilinears $\bar{u}(\mathbf{p}')\Gamma_i u(\mathbf{p})$ ($\Gamma_i = 1, \gamma_0, \gamma_1, \gamma_2, \gamma_3, \gamma_5$) may be shown to behave as constants. However, if a $v$-spinor occurs in the bilinear, as is true in $Z$-graphs, different vertices behave differently (see (Exercise 2.3):

$$
\text{(a)}\quad \left\{
\begin{aligned}
&\bar{v}(\mathbf{p}')\gamma_{0,3}u(\mathbf{p}) \simeq O\left(\frac{1}{P}\right),\\
&\bar{u}(\mathbf{p}')\gamma_{0,3}v(\mathbf{p}) \simeq O\left(\frac{1}{P}\right),
\end{aligned}
\right.
$$

and

$$\text{(b)} \quad \begin{cases} \bar{v}(\mathbf{p}')(1, \gamma_5, \gamma_{1,2})u(\mathbf{p}) \simeq O(P), \\ \bar{u}(\mathbf{p}')(1, \gamma_5, \gamma_{1,2})v(\mathbf{p}) \simeq O(P). \end{cases} \tag{2.20}$$

Thus it is possible for one $Z$-type energy denominator to be compensated by two numerators of type (b), as shown in eqn (2.20). These vertices are denounced as 'bad' vertices. On the other hand, type (a) numerators cause no trouble, and the corresponding vertices are praised as 'good' vertices. Now we note from eqn (I.7) that knowing $W^{\mathrm{e}}_{\mu\nu}$ for $\nu$, $\mu = 0, 3$ is sufficient to determine $W^{\mathrm{e}}_1$ and $W^{\mathrm{e}}_2$. This follows because in the infinite momentum frame chosen via eqns (2.15) and (2.17) we obtain

$$\lim_{P \to \infty} \frac{W^{\mathrm{e}}_{03}}{P^2} = \frac{W^{\mathrm{e}}_2}{M^2} + O\left(\frac{1}{P^2}\right) \tag{2.21a}$$

and

$$\lim_{P \to \infty} W^{e}_{03} - W^{e}_{00} = W^{\mathrm{e}}_1 - \frac{W^{\mathrm{e}}_2 \nu^2}{Q^2} + O\left(\frac{1}{P^2}\right). \tag{2.21b}$$

Hence we can choose $J^{\mathrm{EM}}_{\mu}$ and $J^{\mathrm{EM}}_{\nu}$ to be 'good' currents corresponding to the 'good' vertices with the matrices $\gamma_0$ and $\gamma_3$. Now in a pseudoscalar theory, the only source of a bad vertex is the matrix $\gamma_5$, which corresponds to meson emission, given the basic interaction of that theory. We can show that this alone cannot make bad $Z$-graphs contribute. This is best seen through an illustration. Take some OFPT diagrams (Fig. 28) of order $g^2 e$. These describe the reaction $\mathrm{e} + \mathrm{N} \to \mathrm{e} + \mathrm{N} + 2\pi$, but we have chosen only the subset of diagrams where the electromagnetic current (indicated by a cross) acts on a nucleon. Merely on the basis of energy denominators and the counting of good and bad vertices in Fig. 28, we can make the following statement on the contributions from the different diagrams (see Exercise 2.4):

$$\begin{aligned} (\mathrm{a}_1):(\mathrm{a}_2):(\mathrm{a}_3) &= 1:1:1, \\ (\mathrm{a}_1):(\mathrm{c}_1):(\mathrm{d}_1):(\mathrm{f}_1) &= 1:P^{-2}:P^{-2}:P^{-2}, \\ (\mathrm{a}_2):(\mathrm{b}_2):(\mathrm{c}_2):(\mathrm{d}_2):(\mathrm{e}_2):(\mathrm{f}_2) &= 1:1:P^{-4}:P^{-4}:P^{-2}:P^{-4}, \\ (\mathrm{a}_3):(\mathrm{b}_3):(\mathrm{c}_3):(\mathrm{d}_3):(\mathrm{e}_3):(\mathrm{f}_3) &= 1:1:P^{-4}:P^{-4}:P^{-2}:P^{-4}. \end{aligned} \tag{2.22}$$

Thus the only graphs contributing to the leading terms are $(\mathrm{a}_1)$, $(\mathrm{a}_2)$, $(\mathrm{a}_3)$, $(\mathrm{b}_2)$, and $(\mathrm{b}_3)$—all of which are covered by the kindergarten calculation. The $Z$-graphs $(\mathrm{b}_3)$ and $(\mathrm{b}_2)$ are not troublesome, because in these cases the $Z$-type intermediate states can be absorbed in the undressing and the redressing of the initial and the final nucleon respectively. We notice that,

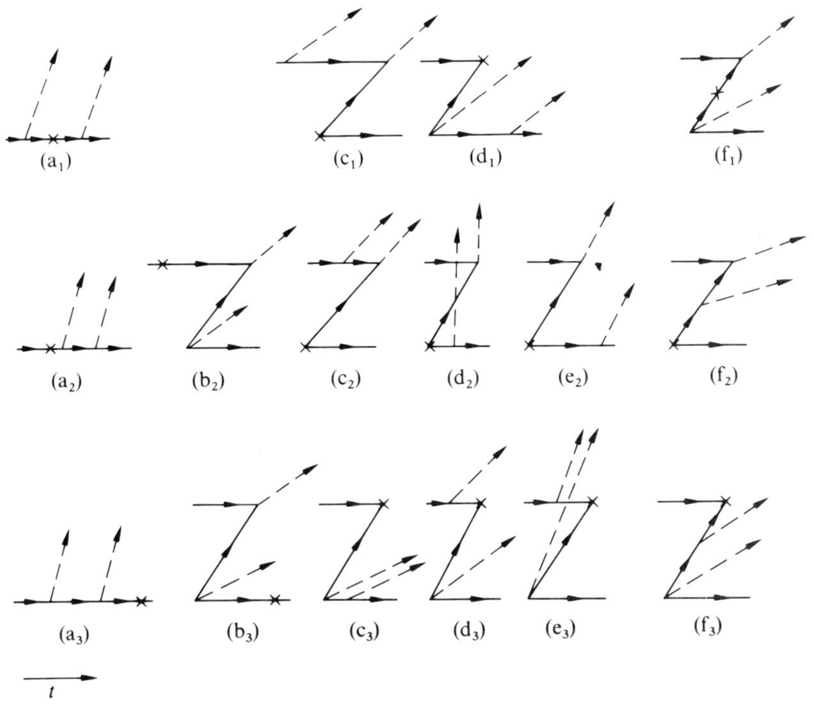


FIG. 28. Old-fashioned perturbation theory diagrams of order $g^2e$ for the reaction $e+N \rightarrow 3+N+2\pi$, where the electromagnetic current (cross) acts only on a nucleon.

had a 'bad' photon vertex been chosen, graphs such as $(c_1)$ and $(d_1)$ would have contributed also.

The above discussion can be extended to any order, and the following general statement can be made. Any final particle belonging to $|Up\rangle$, i.e. one existing just before $j_\nu^{EM}$ acts on it, must go to the right along **P**, otherwise it will enter $|Un\rangle$ still moving to the left (cf. Fig. 28$(f_1)$) and will have at least two bad denominators which cannot be compensated by two bad vertices. (Note that for the leading terms in $P$ the operation of $j_\nu^{EM}$ on $|Up\rangle$ cannot change the direction of the particle belonging to it.) Even though certain $Z$-graphs may contribute to the leading terms, all lines eventually emerging from $|Up\rangle$ move to the right, ensuring $0 < x_i < 1$, so that the result $p^0 - p^0_{Up} \rightarrow 0$ remains valid. Further, since $j_0^{EM}$ and $j_3^{EM}$ introduce negligible longitudinal momenta, all lines also continue to the right in $|Un\rangle$, and hence our proof for $p_n^0 - p_{Un}^0 \rightarrow 0$ goes through also in the deep inelastic limit.

*Final-state interactions*

In the theory of Drell, Levy, and Yan the finiteness constraint on $p_\perp$ is rather critical. This is not only because arguments such as those leading to

$\tau \ll T, T'$ in Chapter 1 need $p_\perp$ to be much less than $\sqrt{Q^2}$. A finite cut-off on $p_\perp$ is also needed in the loop integrations involving virtual partons appearing in the diagram calculations of DLY; otherwise logarithmic violations of scaling behaviour (e.g. a factor of $\ln(Q^2/\mu^2)$) emerge from these integrations.

The presence of an imposed finite cut-off on $p_\perp$, however, has profound consequences on the behaviour of the partons in $|Un\rangle$. In the matrix element $\langle n|U^{-1}j_\nu^{\mathrm{EM}}(0)|Up\rangle$, the state $|Up\rangle$, preceding the action of the bare electromagnetic current $j_\nu^{\mathrm{EM}}(0)$, describes a configuration which involves the emission and re-absorption of bosons and fermion—antifermion pairs. One of the charged constituents in $|Up\rangle$ gets scattered by the bare current and acquires a very large transverse momentum $q_\perp$ of the order of $\sqrt{Q^2}$ (N.B. $q_\parallel \simeq 0$). The unscattered bunch keep moving and emit and re-absorb bosons and fermion–antifermion pairs. They form a group of particles (A in Fig. 29)

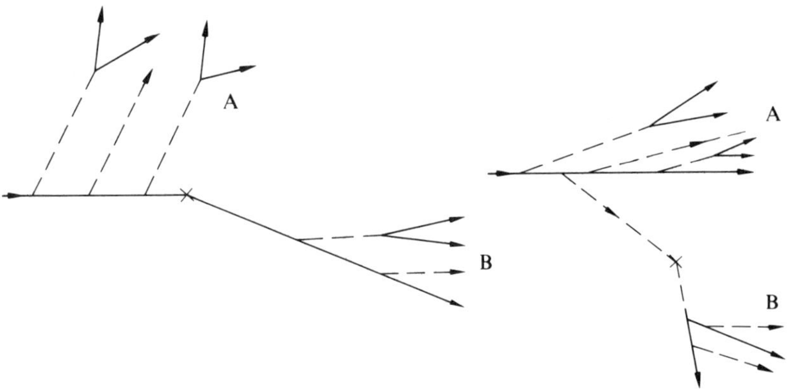


FIG. 29. Non-interfering particle groups (A and B) in the Drell–Levy–Yan theory.

moving very close to each other (i.e. within a small transverse momentum spread) along $\mathbf{P}$. Meanwhile, the scattered charged constituent also emits and re-absorbs bosons and fermion–antifermion pairs which form a second group of particles (B in Fig. 29), again close to each other but along $\mathbf{P}+\mathbf{q}_\perp$ (with $q_\perp \gg p_\perp$). Because of the large difference between the transverse momenta of group A and those associated with B, any final-state interaction between the two groups in the leading term is prevented by the cut-off in $p_\perp$. In other words, in the infinite momentum frame all diagrams with exchanges between A and B go as $(p_\perp/q_\perp)^n$ ($n$ being a positive quantity) relative to the non-interfering diagrams. Typical examples are the diagrams of Fig. 30, whose contributions vanish as $q_\perp \to \infty$. This lack of interference is the original DLY prediction. In the laboratory frame this means that there will be two distinct non-interfering groups A and B; the particles in B moving with very

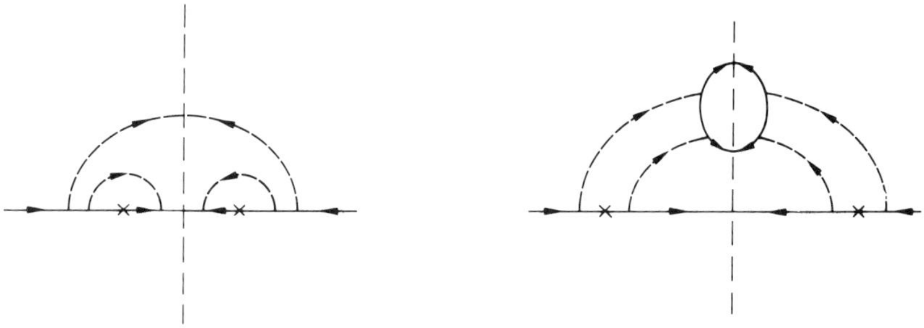

FIG. 30. Typical (asymptotically-vanishing) diagrams for the A–B interference of Fig. 29.

high momenta along **q** leave behind those in A. A number of problems arise in connection with this prediction. The first is that the observation of $R \equiv \sigma_L/\sigma_T \simeq 0$ implies that the scattered partons are predominantly of spin-$\frac{1}{2}$. Hence in group B there will be at least one spin-$\frac{1}{2}$ particle. Using this idea, a pion-nucleon theory, and any reasonable model for the angular distribution of particles in this group around the **q**-direction, the expected yield of nucleons turns out to be considerably larger than the number seen (Berkelman 1971, 1972). The second problem is that no clear indication of a two-jet structure is seen in bubble-chamber studies of deep inelastic hadronic final states in $\mu p$ scattering (Bloom 1972). Finally, if the partons are quarks and gluons and if there are no actual quarks in the final states, what happens to the fractional charge of group B?

In the light of the above discussion, if we want to have quark–partons, we can take two attitudes.

1. The result of no interference between the groups A and B—derived by DLY in a pseudoscalar-meson-baryon theory—is peculiar to that specific interaction and not true in general; in fact, it has been claimed recently (Jackiw and Waltz 1972) that with a quark–vector-gluon theory, defined by an interaction $\mathscr{L}_I = g\bar{q}\gamma^\mu q B_\mu$, exchanges between the two final groups cannot be ignored in the scaling region.

2. The DLY result, true order by order, gets slightly modified when all the orders are summed. For example there may be a final-state interaction term proportional to $(\ln Q^2/\mu^2)^n/n!Q^2$, in an $n$th-order graph, which individually vanishes when $q_\perp$ goes to infinity; but the total sum $\sum_n (\ln Q^2/\mu^2)^n/(n!Q^2) = \mu^{-2}$ implies a constant contribution from final-state interaction. Such a possibility had been investigated mathematically in a model field theory (Chang and Yan 1971), however, some authors (Berman, Bjorken, and Kogut 1971; Bjorken 1971a) have also suggested an intuitive physical picture wherein it may be realized by means of a straggling mechanism. If the scattered quark is unable to emerge, it keeps emitting jets of gluons or quark–antiquark pairs leading to hadrons, until it loses all its momentum

(in *all* directions) and becomes a slow, or 'wee', quark. By that time it no longer has any large transverse momentum difference with group A, and combines with an antiquark in that group. We can visualize a mechanism that is somewhat analogous to the straggling of an electron which stops in an emulsion. Another way of looking at it is to invoke the polarizability of the vacuum and picture the backflow of the quark charge in terms of a polarization current. In any event, all this is assumed not to affect the leading terms of $W^{e}_{\mu\nu}$, so that the standard parton calculations go through. Only the 'signal' of the quark (i.e. its fractional charge) gets lost. In addition, the clear two-jet structure of DLY presumably gets washed out. To understand this point more clearly, let $x$ be the fraction of the longitudinal momentum belonging to the scattered constituent and $y$ the fraction of the total momentum of group B (along $x\mathbf{P}+\mathbf{q}_\perp$) carried by the straggler. The momentum of the latter then (with $k_\perp = O(p_\perp)$) is

$$y(x\mathbf{P}+\mathbf{q}_\perp)+\mathbf{k}_\perp = \frac{yx}{1-x}(1-x)\mathbf{P}+y\mathbf{q}_\perp+\mathbf{k}_\perp .$$

Thus, if $y \simeq mq_\perp^{-1}$ as $q_\perp \to \infty$, then the quark no longer has any large transverse momentum compared to group A, which has a total momentum nearly equal to $(1-x)\mathbf{P}$. The reader is cautioned against accepting too readily this type of mechanism for understanding the non-appearance of real quarks among the final states, since it may violate conventional ideas of short-range forces and of short-range correlations in multi-particle production. The problem of providing a realistic description of deep inelastic final states on the basis of parton considerations is a complex one, but some advances have been made along this line (Bjorken 1971; Feynman 1972; Yan 1973). However, we shall not pursue this problem any further here.

### *OFPT diagram calculation*

We now provide the reader with a typical diagram calculation of the DLY model in a pseudoscalar pion–nucleon theory. Take the infinite momentum frame, where

$$p^\mu = \left(P+\frac{M^2}{2P}, 0, 0, P\right),$$

$$q^\mu = \left(\frac{2M\nu-Q^2}{4P}, \mathbf{q}_\perp, \frac{-2M\nu-Q^2}{4P}\right).$$

To order $ge$ for the cases where the current lands on a nucleon, we see, following our previous arguments, that only the diagram of Fig. 31 will contribute to the leading terms in the scaling region.

Choose

$$\mathbf{P}_1 = x\mathbf{P}+\mathbf{p}_\perp, \qquad \mathbf{k} = (1-x)\mathbf{P}-\mathbf{p}_\perp,$$

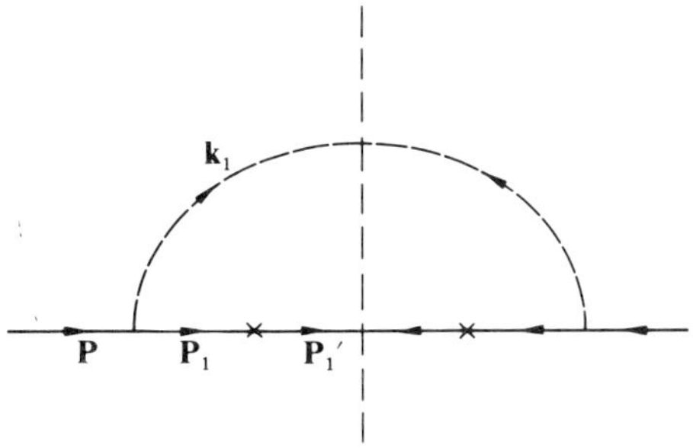


FIG. 31. Diagram for the leading terms in the scaling region.

where $0 < x < 1$ and $\mathbf{p}_\perp \cdot \mathbf{P} = 0$. Momentum conservation implies $\mathbf{P}_1' = \mathbf{P}_1 + \mathbf{q}$. By using OFPT, we may write

$$W^{\mathrm{e}}_{\mu\nu} = \frac{g^2}{(2\pi)^3}\frac{1}{2M}\int\frac{\mathrm{d}^3k_1}{2k_1^0}\frac{1}{2P_1^0}\delta(q^0+p^0-P_1'^0-k_1^0)\times$$
$$\times\tfrac{1}{2}\,\mathrm{Tr}\{(\not{P}_1'+M)\gamma_\mu(\not{P}_1+M)\gamma_5(\not{p}+M)(\not{P}_1-M)\gamma_5\gamma_\nu]\times$$
$$\times(2P_1^0)^{-2}(p^0-P_1^0-k_1^0)^{-2}, \tag{2.23}$$

where $\mu, \nu = 0, 3$. For $0 < x < 1$, we have

$$q^0+p^0-P_1^0-k_1^0$$
$$\simeq \frac{2M\nu-Q^2}{4P}+P+\frac{M^2}{2P}-\left\{xP-\frac{2M\nu+Q^2}{4P}+\frac{(\mathbf{q}_\perp+\mathbf{p}_\perp)^2}{2xP}\right\}-$$
$$-\left\{(1-x)P+\frac{k_\perp^2+\mu^2}{2(1-x)P}\right\}$$
$$\simeq\frac{M\nu}{P}-\frac{q_\perp^2}{2xP}$$
$$\simeq q_0+P_1^0-P_1'^0. \tag{2.24}$$

Eqn (2.24) demonstrates the near-conservation of energy across the electromagnetic vertex in the infinite momentum frame. Further, we may write

$$\delta(q^0+p^0-P_1^0-k_1^0)\simeq\frac{2Pw^2}{Q^2}\delta(x-w). \tag{2.25}$$

Moreover, the energy denominator can be rewritten as

$$2P_1^0(p^0-P_1^0-k_1^0)\simeq-\frac{1}{1-x}\{k_\perp^2+M^2(1-x)^2+\mu^2x\}. \tag{2.26}$$

Using eqns (2.24) and (2.25) in eqn (2.23), substituting $P_1^\mu \simeq xp^\mu, P_1'^\mu \simeq xp^\mu + q^\mu$ in its numerator in the leading order, and comparing with eqns (I.7) and (I.32), we obtain

$$\lim_{\mathrm{Bj}} MW_1^{\mathrm{e}} = F_1^{\mathrm{e}} = F_2^{\mathrm{e}}/2w,$$

$$\begin{aligned} F_2^{\mathrm{e}} &= \lim_{\mathrm{Bj}} \nu W_2^{\mathrm{e}} \\ &= \frac{g^2}{16\pi^2} w(1-w) \int_0^{\lambda^2} \mathrm{d}k_\perp^2 \frac{k_\perp^2 + M^2(1-w)^2}{\{k_\perp^2 + M^2(1-w)^2 + \mu^2 w\}^2} \\ &= \frac{g^2 w(1-w)}{16\pi^2}\left[\ln\left\{1 + \frac{\lambda^2}{M^2(1-w)^2+\mu^2 w}\right\} - \right. \\ &\left. - \frac{\mu^2 w \lambda^2}{\{\lambda^2 + M^2(1-w)^2 + \mu^2 w\}\{M^2(1-w)^2 + \mu^2 w\}}\right]. \end{aligned} \tag{2.27}$$

In obtaining eqn (2.27), we have replaced the phase integral $\int \mathrm{d}^3k$ by $2\pi P \times \int_0^1 \mathrm{d}x \int_0^{\lambda^2} \mathrm{d}k_\perp^2$, where $\lambda$ is the finite cut-off. The $x$-integration has then been disposed of by the $\delta(x-w)$ shown in eqn (2.25). Eqn (2.27) shows Bjorken scaling explicitly (for this diagram) provided $\lambda$ is finite. A similar calculation can be done (see Exercise 2.5) for the same diagram as Fig. 30, but with the electromagnetic current landing on the pion.

In the above discussion we have shown how the DLY model justifies the basic parton picture for deep inelastic lepton–hadron scattering. However, since the model is quite powerful and has such features as crossing symmetry, it can be applied to other processes like $\mathrm{e}^+ + \mathrm{e}^- \to \mathrm{H} +$ 'anything', $\mathrm{H} + \mathrm{N} \to l^+ + l^- +$ 'anything', and $l + \mathrm{N} \to l' + \mathrm{H} +$ 'anything' (Drell and Yan 1971). Before leaving this topic, we should also say a few general words about field-theoretical approaches to the problem of Bjorken scaling in deep inelastic lepton–hadron scattering. Any regular order-by-order perturbation approach to this problem, based on diagrammatic calculations with a renormalizable interaction, yields logarithmic violations of Bjorken scaling (Christ 1972). The DLY scheme evades this feature by the arbitrary imposition of a transverse momentum cut-off in the loop integrals. Moreover, present SLAC data up to $Q^2 = 8\ \mathrm{GeV}^2$ and $\nu = 20$ GeV can be made to accommodate small logarithmic departures from scaling. Hence one can legitimately take the attitude that the perturbative predictions of such scaling violations will be vindicated once data for larger values of $Q^2$ and $\nu$ become available. Another view is that the scaling property should be regarded as valid only in an intermediate energy range. The familiar GeV scale, where it works is not of asymptotic significance, but the energy range needed for its breakdown is as yet unreached. In fact, a parton-like model embodying the latter attitude has been given (Drell and Lee 1972, Lee 1972).

Here, the nucleon is regarded as a bound state of a bare (spin-$\frac{1}{2}$) fermion and a bare pseudoscalar meson, which exchange a scalar quantum. The virtual $\gamma^*$ can scatter from either constituent. The model is covariant and partially non-perturbative in the sense of having summed ladder diagrams like those of Bethe and Salpeter. An effective point-like $\gamma^*$–parton vertex is 'derived', leading to Bjorken scaling in the range $\nu/M = O(1)$. However, owing to processes such as the emission and absorption of hard mesons by the fermion, scaling violations emerge around $\nu/M = O(e^{1/\varepsilon})$, where $\varepsilon \equiv f^2/4\pi$ ($f$ being the fermion scalar-meson coupling constant) is expected to be $O(10^{-1})$.

### The model of Landshoff, Polkinghorne, and Short

A non-perturbative approach to the parton picture (the LPS model), formulated by the authors named above, has been successful in obtaining the scaling behaviour for the deep inelastic structure functions of lepton–nucleon scattering (Landshoff and Polkinghorne 1972). This model has already been compared with the DLY theory, and we shall not repeat those points here. It is sufficient to focus on the non-perturbative aspect of the approach and on its formal lack of a field-theoretical Lagrangian. Although Bjorken scaling emerges in both models, their predictions for all deep inelastic processes are not the same. The difference between the two approaches should be of interest, therefore, in the context of current and future experimental results. We give here the gist of the LPS model, confining ourselves for the sake of brevity, to electron–proton scattering; the generalization to neutrinos and antineutrinos is fairly straightforward.

The tensor $W^{\mathrm{e}}_{\mu\nu}$, introduced on p. 4, is the imaginary part of $T^{\mathrm{e}}_{\mu\nu}$. The latter is proportional to the forward off-shell Compton amplitude, which is defined to be the continuation off the photon's mass shell of the connected part of the matrix-element $\langle \gamma p_{\mathrm{out}} | \gamma p_{\mathrm{in}} \rangle$ averaged over proton-spins. If this matrix element is written as $e^2(2\pi)^{-3}(2q_0)^{-1}\varepsilon^{*\mu}\varepsilon^{\nu}T^{\mathrm{e}}_{\mu\nu}$, where $\varepsilon$ refers to the photon's polarization, then we have

$$e^2 T^{\mathrm{e}}_{\mu\nu} = (2\pi)^2 \frac{p_0}{M} 2\mathrm{i} \int \mathrm{d}^4x\, \mathrm{e}^{\mathrm{i}q.x} \lim_{y\to 0} \Box_x \Box_y \langle p | T A_\mu(x) A_\nu(0) | p \rangle. \quad (2.28)$$

In eqn (2.28) the photon field $A_\mu$ appears through the use of the standard reduction technique (Gasiorowicz 1966) and is to be understood as a Heisenberg field. It follows from eqn (2.28) that

$$T^{\mathrm{e}}_{\mu\nu} = (2\pi)^2 \frac{p_0}{M} 2\mathrm{i} \int \mathrm{d}^4x\, \mathrm{e}^{\mathrm{i}q.x} \langle p | T J^{\mathrm{EM}}_\mu(x) J^{\mathrm{EM}}_\nu(0) | p \rangle +$$

$$+ \text{tensors with real polynomial coefficients.} \quad (2.29)$$

Taking the imaginary part of eqn (2.29) and comparing with eqn (I.10), we obtain

$$\operatorname{Im} T^{e}_{\mu\nu} = W^{e}_{\mu\nu}. \tag{2.30}$$

We thus have to deal essentially with the matrix element

$$M^{e}_{\mu\nu} = (2\pi)^2 \frac{p_0}{M} 2\mathrm{i} \int \mathrm{d}^4x\, \mathrm{e}^{\mathrm{i}q.x} \langle p| T J^{\mathrm{EM}}_{\mu}(x) J^{\mathrm{EM}}_{\nu}(0)|p\rangle. \tag{2.31}$$

The partons are introduced by considering the electromagnetic current as a parton current. The photon vertex is considered as a point vertex, and the parton field is taken either as a scalar field $\phi$, with the current $J_\mu = \mathrm{i}\phi^* \overleftrightarrow{\partial}_\mu \phi$, or as a spin-$\frac{1}{2}$ field $\psi$ with the current $J_\mu = \bar{\psi}\gamma_\mu\psi$.

*Spin-0 case*

Substituting the spin-0 form of the parton current in $M^{e}_{\mu\nu}$ and using the standard reduction technique, we obtain

$$M^{e}_{\mu\nu} = \frac{2\mathrm{i}}{(2\pi)^6} \int \mathrm{d}^4k_1 \int \mathrm{d}^4k_2 (2k_1 - q)^\mu (2k_2 - q)^\nu \times$$
$$\times T_6(k_1 - q, k_1, k_2, k_2 - q). \tag{2.32}$$

In eqn (2.32) the six-point function $T_6$, which couples four virtual partons (each current contributing two) to two spin-averaged protons, is 'non-amputated,' i.e. it includes propagators for its parton legs. It can be decomposed into three parts, as shown diagrammatically in Fig. 32. These

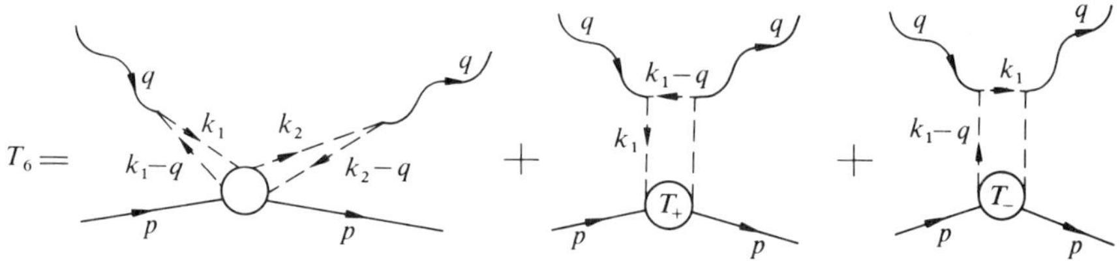


FIG. 32. Three-part decomposition of the six-point function $T_6$ of eqn (2.32).

diagrams are not Feynman graphs; each blob is the connected part of a complete amplitude in the LSZ sense (Gasiorowicz 1966). We see that $T_6$ contains a purely connected part describing the scattering of three particles (parton, antiparton, and proton) into themselves. In addition, there are partially disconnected parts (Fubini and Furlan 1968) in which either of the partons is a mere spectator. The arrows depict the flow of the labelled momenta and also distinguish partons from antipartons. The contribution to $T_6$ from the disconnected parts is

$$T^{D}_{6} = (2\pi)^4 \delta^{(4)}(k_1 - k_2)[T_+ \Delta_{\mathrm{F}}\{(k_1 - q)^2\} + T_- \Delta_{\mathrm{F}}(k_1^2)], \tag{2.33}$$

where $T_{\pm}$ are the amplitudes for the two-body forward scattering of virtual partons and antipartons respectively from a spin-averaged proton and $\Delta_F(k^2)$ is the parton propagator

$$\Delta_F(k^2) = i\int d^4x\, e^{ik.x}\langle 0|T\phi^*(x)\phi(0)|0\rangle.$$

We then require that, for large $k^2$, $\Delta_F(k^2)$ have the asymptotic form $(k^2)^{-1}$ of a free scalar propagator. Moreover, the softness hypothesis is introduced in the form of the dictum that hadronic amplitudes with virtual partons among external legs must go 'sufficiently' rapidly to zero as the masses of the partons become large. This postulate is a questionable one but is essential to the analysis of LPS, which is therefore also called the softened-field parton model. It has the immediate consequence that the partially disconnected parts in $T_6$ dominate in the deep inelastic limit. This occurs for the following reason. The damping of the connected parton amplitudes due to increasing virtual masses of the partons is far more severe than the $1/k^2$ fall-off of the parton propagator. Hence the piece with the four connected external parton legs loses out asymptotically.

We now explore the deep inelastic limit of $M^{e}_{\mu\nu}$, following LPS. It is convenient for this purpose to introduce two variables $x$, $y$ via the decomposition

$$k = xp + yq + \kappa, \tag{2.34}$$

where $\kappa$ is a two-dimensional space-like vector orthogonal to $p$ and $q$. First, take the right-most of the diagrams in Fig. 32. In the deep inelastic limit the virtual mass squared $\mu^2$ of the partons connected to the blob is

$$\mu^2 = 2M\nu(y-1)\{x - w(y-1)\} + x^2M^2 + \kappa^2. \tag{2.35}$$

The implication of the softness postulate is that, as $\nu \to \infty$, the dominant contributions to the integral in eqn (2.32) can come only from three regions: (1) $y \simeq 1$, $x$, and $\kappa^2$ finite, (2) $y \simeq 1+(x/w)$, $x$ and $\kappa^2$ finite, and (3) $x$ and $y$ finite but arbitrary and $\kappa^2/\nu \simeq -2M(y-1)\{x - w(y-1)\}$.

Consider the said regions one by one, taking case (1) first. For this define $\bar{y}$ by

$$\bar{y} \equiv 2M\nu(y-1),$$

so that the parton's virtual squared mass can be approximated to be $\mu^2 \simeq x\bar{y} + x^2M^2 + \kappa^2$. Moreover, from eqn (2.34), it follows that

$$k^2 \simeq M\nu(x-w), \tag{2.36a}$$

$$s \equiv (p+q-k)^2 \simeq (x-1)\bar{y} + (x-1)^2M^2 + \kappa^2. \tag{2.36b}$$

Let us recast the integration of eqn (2.32) in terms of the variables $x$, $\bar{y}$, and $\kappa$ instead of $k$. We can use the information of eqn (2.33) as well as the respective behaviour of $\Delta_F(k)$ and of $y(\simeq 1/k^2$ and unity respectively in the limit under consideration) and finally eqn (2.35a) in eqn (2.32) to obtain the contribution to $M^e_{\mu\nu}$ as

$$\frac{i}{(2\pi)^2 M\nu} \int_{-\infty}^{\infty} dx \int_{-\infty}^{\infty} d\bar{y} \int d^2\kappa \, \frac{(2xp+q+2\kappa)_\mu (2xp+q+2\kappa)_\nu T_-(s, \mu^2)}{x-w}, \tag{2.36c}$$

with $\mu^2$ as defined in eqn (2.33). In addition to having singularities in $\mu^2$, $T_-$ is seen, from analyticity considerations, to possess a 'right-hand' cut in $s$ and a 'left-hand' cut in

$$u = 2\mu^2 + 2M^2 - s \simeq (x+1)\bar{y} + (x+1)M^2 + \kappa^2, \tag{2.37}$$

with the singularities being understood as located slightly below the positive real axis of the relevant variable. We now examine the $\bar{y}$ integration. If $|x| > 1$, all the singularities are on the same side of the real axis in the $\bar{y}$-plane, so that the completion of the contour by a semicircle at infinity in the opposite half-plane gives a null result. On the other hand, for $0 < x < 1$, the integral is non-zero but—by virtue of the permitted addition of a semicircle at the upper-half infinity—may be taken around the $s$-cut, i.e. around those singularities which appear slightly above the real $\bar{y}$-axis. Similarly, for $-1 < x < 0$, the $\bar{y}$ integration, aided by the permissible inclusion of an additional semicircle at the lower-half infinity, may be taken around the $u$-cut, i.e. around those singularities which are located slightly below the real $\bar{y}$-axis. Finally, $T_-$ in eqn (2.36c) may be replaced by

$$2i\{\theta(x)\theta(1-x) \operatorname{Im} T_{R-} + \theta(-x)\theta(1+x) \operatorname{Im} T_{L-}\},$$

where the subscripts L, R refer respectively to the 'right' and 'left' parts of the discontinuity in $T_-$. In the physical situation, with $0 < w < 1$, the expression in eqn (2.36c) develops an imaginary part obtained by replacing the denominator of the integral by $\delta(x-w)$. As a result, only the $T_R$ piece survives. The other partially disconnected diagram can be handled similarly, or the corresponding result may be obtained by crossing. In this case only the 'left' part, involving $T_L$, contributes. Finally, adding the two and isolating the coefficient of the $p^\mu p^\nu$ term (cf. eqn (I.7), the $p^\mu p^\nu$ part coming from the $\kappa^\mu \kappa^\nu$ term fails to survive in the deep inelastic limit), we obtain

$$F^e_2(w) = \lim_{\mathrm{Bj}} \nu W^e_2 = -\frac{2w^2}{\pi M} \int_{-\infty}^{\infty} dy \int d^2\kappa$$

$$\{\operatorname{Im} T_{R-}(s, \mu^2) + \operatorname{Im} T_{L+}(u, \mu^2)\}, \tag{2.38}$$

where

$$\begin{aligned} \mu^2 &= \bar{y}w + w^2M^2 + \kappa^2, \\ s &= -(1-w)\bar{y} + (1-w)^2M^2 + \kappa^2, \\ u &= (1+w)\bar{y} + (1+w)^2M^2 + \kappa^2. \end{aligned} \tag{2.39}$$

The second region (case (2)) can only give a subdominant contribution in the limit of interest. This is best seen by defining $\bar{\bar{y}} \equiv 2M\nu(y-1-xw^{-1})$. Then $s \simeq -2xw^{-1}\nu$, so that $\bar{\bar{y}}$ appears in $T_-(s, \mu^2)$ only through its dependence on $\mu^2$. As a result, the corresponding integration can be completed by a semicircle at infinity and goes to zero as $\nu \to \infty$, the vanishing rate depending on the attenuation of $T_-$ in the limit of large $\mu^2$ and $s$. Once again, by use of crossing, the corresponding contribution from the other partially disconnected diagram also may be ignored. Finally, we may dispose of the region referred to in case (3) in a similar fashion. Hence eqn (2.38) is the end result for $\nu W_2^e$ in the deep inelastic limit. It manifestly displays Bjorken scaling. Moreover, since a $g_{\mu\nu}$ term (cf. eqn (I.7)) is seen not to emerge in $M_{\mu\nu}^e$ from the above considerations, we must conclude that

$$F_1^e(w) = \lim_{\text{Bj}} MW_1^e = 0.$$

*Spin-$\frac{1}{2}$ case*

The analysis here proceeds broadly along the same path as that outlined above. The major difference is that the parton propagator is now

$$S_F(k) = \mathrm{i}\int \mathrm{d}^4x\, \mathrm{e}^{\mathrm{i}k.x}\langle 0|T\bar{\psi}(x)\psi(0)|0\rangle.$$

The requirement on this is that for large $k^2$ it behave as $\not{k}/k^2$—the form for a free spin-$\frac{1}{2}$ propagator. The contribution of interest to $M_{\mu\nu}^e$ in this case, which would be analogous to that of eqn (2.36) for spin-0 partons, is

$$\frac{\mathrm{i}}{2(2\pi)^2M\nu}\int_{-\infty}^{\infty}\mathrm{d}x\int_{-\infty}^{\infty}\int \mathrm{d}\bar{y}\,\mathrm{d}^2\kappa\,\frac{\mathrm{Tr}\{\gamma_\mu(x\not{p}+\not{q}+\not{\kappa})\gamma_\nu T_-\}}{x-w}. \tag{2.40}$$

In the expression given by eqn (2.40) $T_-$ is a matrix in the parton spinor space. Since $p$ and $(k-q)$ are the only two four-vectors on which $T_-$ can depend, its contribution to the trace in eqn (2.40) may be taken as

$$T_1^-\not{p} + T_2^-(\not{k}-\not{q}) = (T_1^- + xT_2^-)\not{p} + T_2^-\not{\kappa},$$

where the $T_{1,2}^- \equiv T_{1,2}^-(s, \mu^2)$ but are otherwise unknown. The trace may be written as

$$4\{2xp_\mu p_\nu(T_1^- + xT_2^-) + (p_\mu q_\nu + p_\nu q_\mu)(T_1^- + xT_2^-) - p.q(T_1^- + xT_2^-)g_{\mu\nu} + \\ + \text{non-leading terms}\}.$$

Hence the coefficients of the $-g^{\mu\nu}$ and of the $p^\mu p^\nu$ terms are in the ratio $M\nu/2x$, leading to the expected result

$$\lim_{\mathrm{Bj}} MW_1^{\mathrm{e}} = \frac{\lim_{\mathrm{Bj}} \nu W_2^{\mathrm{e}}}{2w}. \tag{2.41}$$

Moreover, carrying through the same analysis as in the earlier case, we obtain

$$\begin{aligned}\lim_{\mathrm{Bj}} \nu W_2^{\mathrm{e}} &= F_2^{\mathrm{e}}(\omega)\\ &= -\frac{2w}{\pi M}\int_{-\infty}^{\infty} \mathrm{d}\bar{y} \int \mathrm{d}^2\kappa\, [\mathrm{Im}\, T_{1\mathrm{R}}^{-}(s,\mu^2) + \mathrm{Im}\, T_{1\mathrm{L}}^{+}(s,\mu^2)\\ &\qquad + w\{\mathrm{Im}\, T_{2\mathrm{R}}^{-}(s,\mu^2) + \mathrm{Im}\, T_{2\mathrm{L}}^{+}(s,\mu^2)\}].\end{aligned} \tag{2.42}$$

The scaling manifest in eqn (2.42) and the relation of eqn (2.41) again agree with the canonical results for spin-$\frac{1}{2}$ partons obtained by the kindergarten methods of Chapter 1.

## Exercises

2.1. Derive eqn (2.9), in which the physical proton is 'undressed'.

2.2. Verify the unitarity condition eqn (2.11) to the lowest non-trivial order.

2.3. In the infinite momentum frame demonstrate the behaviour of type (a), or 'good', and of type (b), or 'bad', vertices as given in eqn (2.20).

2.4. Show that in the infinite momentum frame the contributions from the different diagrams of Fig. 28 behave as stated in eqn (2.22).

2.5. Verify the results of eqns (2.27). Do the corresponding calculation for the same diagram (Fig. 31), but with the electromagnetic current landing on the pion.

# PART II
# SCALE INVARIANCE

# 3

## FUNDAMENTALS OF SCALE INVARIANCE

### General concepts

THE relevance of the notion of scale or dilation symmetry to deep inelastic lepton–hadron processes became clear after the observation of Bjorken scaling in the SLAC electro-production data. However, the concept of scale invariance as a possible symmetry of the fundamental interactions was introduced by Kastrup long before that. In any event, interest in the subject had been revived (Wilson 1969; Gell-Man 1969) in the late 1960s in the context of connections between dilation invariance and current algebra. In the next chapter we shall discuss how Bjorken scaling relates to scale invariance, but first let us consider the basic ideas (Mack 1967; Carruthers 1971; Jackiw 1971*a*, 1972*b*) on which that logic is founded.

*Scale dimensions*

Under a scale transformation or dilation the coordinates change by an over-all positive factor

$$x \to x' = e^{\varepsilon}x, \tag{3.1}$$

where $\varepsilon$ is a real number. Scale invariance refers to the symmetry of the group of scale transformations. This is a space–time symmetry based on kinematic transformations as opposed to internal symmetries such as SU(3) (Gell-Mann and Nee'man 1964), which are based on gauge transformations. The simplest behaviour of a field operator $\phi(x)$ under a scale transformation is to be changed linearly, i.e.

$$\phi(x) \to \phi'(x) = e^{\varepsilon d}\phi(e^{\varepsilon}x), \tag{3.2}$$

where $d$ is a parameter known as the scale dimension of the field. Mention should be made here of the existence of field theories where scale transformations act non-linearly. For example, we may define a new field $\kappa$ by

$$f^{-1}\,e^{f\kappa} = \phi,$$

where $f$ is a dimensional constant. Then, under a scale transformation, $\kappa$ changes in the following non-linear way:

$$\kappa(x) \to \kappa'(x) = \kappa(e^{\varepsilon}x) + d\varepsilon f^{-1}.$$

Thus a scale dimension cannot be defined for $\kappa(x)$. This type of field is a common occurrence in theories with non-polynomial Lagrangians. However, we shall be concerned with theories in which the transformation of any field under dilation is linear. Let us consider a scalar field $\phi$ and a spin-$\frac{1}{2}$ field $\psi$. For free fields $\phi_0, \psi_0$ or unrenormalized interacting fields $\phi_u, \psi_u$, the equal-time commutation relations

$$[\phi(x), \dot{\phi}(y)]_{x^0=y^0} = i\delta^{(3)}(\mathbf{x}-\mathbf{y}), \tag{3.3a}$$

$$\{\psi(x), \psi^{\dagger}(x)\}_{x^0=y^0} = \delta^{(3)}(\mathbf{x}-\mathbf{y}) \tag{3.3b}$$

imply that $d_\phi = 1$ and $d_\psi = \frac{3}{2}$. These numbers stand for the canonical scale dimensions of the corresponding fields. But we should emphasize that, in the presence of interactions, eqns (3.3) may no longer be valid for the fields of interest; thus these simple dimensional values are not necessarily right. For renormalized fields we have to consider $\phi_R = (\sqrt{Z})^{-1}\phi_u$, where $Z$ is the renormalization constant. Then the above analysis fails if $Z$ is 0.

At this stage it is important to emphasize the difference between scale dimensions and the ordinary dimensions of elementary physics. The latter are understood in terms of fixed dimensional constants—masses and dimensional couplings, for instance—whose relative values fix the scale of length. The ordinary dimensions of several quantities in conformity with our convention are listed in Table 3.1. Scale dimensions, in contrast, are defined

TABLE 3.1

| Quantity | Ordinary dimension | Remarks |
|---|---|---|
| Mass $M$ | 1 | |
| Time $T$ | $-1$ | $\hbar = 1$ |
| Length $L$ | $-1$ | $c = 1$ |
| Action $= \int d^4x\mathscr{L}$ | 0 | |
| Lagrangian density $\mathscr{L}$ | 4 | |
| Scalar field $\phi$ | 1 | From the $m^2\phi^2$ term in $\mathscr{L}$ |
| Spin-$\frac{1}{2}$ field $\psi$ | $\frac{3}{2}$ | From the $m\bar{\psi}\psi$ term in $\mathscr{L}$ |

only for certain field operators which are among the dynamical variables of a physical theory. The coincidence of the canonical scale dimensions of fields with their ordinary dimensions should not be misinterpreted. It does not imply that methods employing scale invariance are mere dimensional analyses with formal trimmings. The transformations of dimensional analysis—by virtue of the consequent changes in masses and certain couplings—turn one physical theory into another. They are always exact symmetries—in the sense of yielding exact solutions to the transformed theory, given such solutions to the original one. On the other hand, scale transformations stay within a given physical theory and their symmetry may be broken, as will be shown shortly. To illustrate the difference between ordinary and scale dimensions note that the term $m^2\phi^2$ has an ordinary dimension of 4 but a canonical scale dimension of 2. Further, the field $\kappa$—introduced previously—has the ordinary dimension of unity, but its scale dimension is not defined. Finally, if equal-time commutation relations are destroyed by renormalization, induced by the interactions, scale dimensions will, in general, change from the canonical numbers as functions of appropriate coupling constants; ordinary dimensions of fields however, will, remain firmly fixed at their ordinary values.

*Consequences of exact scale invariance*

Take a world in which scale invariance is an exact symmetry. Let us consider translations, infinitesimal dilations and homogeneous Lorentz rotations respectively:

$$x^\mu \to x^\mu + a^\mu; \qquad U(a) = \exp(\mathrm{i}a^\mu p_\mu). \tag{3.4a}$$

$$x^\mu \to (1+\varepsilon)x^\mu; \qquad U(D) = \exp(\mathrm{i}\varepsilon D). \tag{3.4b}$$

$$x^\mu \to (g^{\mu\nu} + \alpha^{\mu\nu})x_\nu; \qquad U(\Lambda) = \exp(\mathrm{i}\alpha^{\mu\nu} M_{\mu\nu}). \tag{3.4c}$$

Both the explicit coordinate transformations and the unitary operators effecting these transformations have been listed above. $P_\mu$, $D$, and $M_{\mu\nu}$ are the corresponding generators. We now compare two sequences of dilation and translation, i.e.

$$x^\mu \underset{\varepsilon}{\to} (1+\varepsilon)x^\mu \to (1+\varepsilon)x^\mu + a^\mu,$$

with

$$x^\mu \underset{a}{\to} x^\mu + a^\mu \underset{\varepsilon}{\to} (1+\varepsilon)x^\mu + (1+\varepsilon)a^\mu,$$

to find that in an exactly scale-invariant world we must have

$$U(D)U(a) = U\{(1+\varepsilon)a\}U(D),$$

or

$$\mathrm{e}^{\mathrm{i}\varepsilon D}\,\mathrm{e}^{\mathrm{i}aP} = \mathrm{e}^{\mathrm{i}(1+\varepsilon)a.P}\,\mathrm{e}^{\mathrm{i}\varepsilon D}.$$

The last relation implies

$$\mathrm{i}[D, P_\mu] = P_\mu, \tag{3.5}$$

or

$$[D, P^2] = -2\mathrm{i}P^2. \tag{3.6}$$

Considering two sequences of dilation and homogeneous Lorentz rotation, we find similarly that

$$U(\Lambda)U(D) = U(D)U(\Lambda),$$

which leads to

$$[M_{\mu\nu}, D] = 0 \tag{3.7}$$

in a scale-invariant world. Thus, in the limit of exact scale invariance, $D$ is a Lorentz scalar.

Let us now consider the effect of an infinitesimal dilation on a field operator $\phi(x)$. To lowest order in $\varepsilon$, the consequent change in $\phi$, from eqn (3.2), is

$$\delta\phi \equiv \phi'(x) - \phi(x) = \varepsilon(d + x\,.\,\partial)\phi(x) + O(\varepsilon^2). \tag{3.8}$$

On the other hand, we want the unitary transformation $U(D) = \mathrm{e}^{\mathrm{i}\varepsilon D}$ to have the property that

$$U(D)\phi(x)U^{-1}(D) = \phi'(x) = \mathrm{e}^{\varepsilon d}\phi(\mathrm{e}^{\varepsilon}x).$$

Therefore we can write

$$\delta\phi = \mathrm{i}\varepsilon[D, \phi] + O(\varepsilon^2). \tag{3.9}$$

Comparing eqns (3.8) and (3.9), we have

$$[D, \phi(x)] = -\mathrm{i}(d + x\,.\,\partial)\phi(x). \tag{3.10}$$

Eqn (3.10) is an algebraic relation defining the local transformation property of $\phi(x)$ *vis a vis* the generator of dilation. If scale dimensions are taken to be canonical, for scalar and spin-$\frac{1}{2}$ fields we shall have respectively

$$[D, \phi(x)] = -\mathrm{i}(1 + x\,.\,\partial)\phi(x), \tag{3.10a}$$

$$[D, \psi(x)] = -\mathrm{i}(\tfrac{3}{2} + x\,.\,\partial)\psi(x). \tag{3.10b}$$

Now if we have a simple scale-invariant Lagrangian density such as

$$\mathscr{L} = \mathrm{i}\bar{\psi}\gamma\,.\,\partial\psi + \tfrac{1}{2}(\partial\phi)\,.\,(\partial\phi) + g\bar{\psi}\gamma_5\psi\phi + \lambda\phi^4 \tag{3.11}$$

for a system of scalar and spin-$\frac{1}{2}$ fields, the change in it due to an infinitesimal dilation can be computed readily, using eqns (3.9) and (3.10a). We obtain

$$\delta\mathscr{L} = \varepsilon(4 + x\,.\,\partial)\mathscr{L} + O(\varepsilon^2). \tag{3.12}$$

In other words, the Lagrangian density has the scale dimension of 4. Eqn (3.12) vanishes, when integrated over all space and time, thus ensuring the scale invariance of the action.

*Role of the stress–energy tensor*

The standard stress–energy tensor of Lagrangian field theory is given by (Bjorken and Drell 1965)

$$\tilde{T}_{\mu\nu} \equiv \pi_\mu \partial_\nu \phi - g_{\mu\nu}\mathscr{L}, \tag{3.13}$$

where $\mathscr{L} \equiv \mathscr{L}(\phi, \dot{\phi}, x)$ and $\pi_\mu \equiv \delta\mathscr{L}/\delta\partial^\mu\phi$ is the canonical four-momentum. $\tilde{T}_{\mu\nu}$ is divergenceless and is symmetric for spin-0 fields. The symmetry, however, is absent for higher-spin fields. Thus, for such fields, an explicit symmetrization has to be performed by adding spin-dependent terms. The procedure for accomplishing this was laid out by Belinfante. We write the Lorentz transformation of fields with spin as

$$[M_{\mu\nu}, \phi^{\alpha\cdots}(x)] = -\mathrm{i}(x_\mu\partial_\nu - x_\nu\partial_\mu + \textstyle\sum_{\mu\nu}^{\alpha\ldots,\beta\ldots})\phi^{\beta\cdots}(x), \tag{3.14}$$

where

$$\textstyle\sum_{\mu\nu} = 0 \qquad \text{for spin-0 fields,}$$

$$\textstyle\sum_{ab}^{\mu\nu} = \frac{1}{2}\sigma_{ab}^{\mu\nu} \qquad \text{for spin-}\tfrac{1}{2}\text{ fields,}$$

$$\textstyle\sum_{\mu\nu}^{\alpha\beta} = \delta_\mu^\alpha\delta_\nu^\beta - \delta_\nu^\alpha\delta_\mu^\beta \qquad \text{for spin-1 fields, etc.}$$

Then the symmetrized Belinfante tensor (Schweber 1961) is

$$T_{\mu\nu} = \pi_\mu\partial_\nu\phi - g_{\mu\nu}\mathscr{L} + \partial^\beta\tfrac{1}{2}\{\pi_\beta \textstyle\sum_{\mu\nu}\phi - \pi_\mu \sum_{\beta\nu}\phi - \pi_\nu \sum_{\beta\mu}\phi\}. \tag{3.15}$$

Given the above $T_{\mu\nu}$, we may obtain the standard relations

$$P_\mu = \int \mathrm{d}^3x\, T_{0\mu}, \tag{3.16a}$$

$$M_{\mu\nu} = \int \mathrm{d}^3x\,(x_\mu T_{\nu 0} - x_\nu T_{\mu 0}). \tag{3.16b}$$

There are some problems with the form of the stress–energy tensor given in eqn (3.16). As an operator, $T_{\mu\nu}$ has been found to behave badly in renormalized perturbation theory (Callan, Coleman, and Jackiw 1970). However, Callan *et al.* have given a new, improved stress–energy tensor $\theta_{\mu\nu}$ which retains the necessary properties of $T_{\mu\nu}$ but is better behaved in renormalized perturbation theory. This is given by

$$\theta_{\mu\nu} = T_{\mu\nu} - \sum_\varphi \tfrac{1}{6}(\partial_\mu\partial_\nu - g_{\mu\nu}\Box)\phi^2. \tag{3.17}$$

In eqn (3.17) the summation is over all scalar fields present in the theory. The following properties of $\theta_{\mu\nu}$ are noteworthy:

(1) $\theta_{\mu\nu} = \theta_{\nu\mu}$,

(2) $\partial^\mu\theta_{\mu\nu} = \partial^\nu\theta_{\mu\nu} = 0$,

(3) $$P_\mu = \int d^3x\theta_{0\mu},$$

(4) $$M_{\mu\nu} = \int d^3x(x_\mu\theta_{\nu 0} - x_\nu\theta_{\mu 0}),$$

(5) $$\int d^3x\theta_{00} = \int d^3xT_{00} = H,$$

(6) $$\langle p|\theta_{00}|p\rangle_{\mathbf{p}=0} = \left(\frac{1}{2\pi}\right)^3 m,$$

where in (6) the matrix element has been taken between states of spin-0 and mass $m$. In addition to the properties listed above, there is an important relation between the stress–energy tensor and the dilation current. We shall now introduce the latter. Starting with the scale invariance of the action integral, we follow the standard field-theoretical procedure of defining the current associated with a symmetry. Since, by eqn (3.12), $\mathscr{L}$ has the scale dimension of 4, under an infinitesimal dilation the change in the Lagrangian density may be written as

$$\mathscr{L}(x) \underset{\varepsilon}{\rightarrow} \mathscr{L}'(x) = (1+\varepsilon)^4\mathscr{L}\{x(1+\varepsilon)\}.$$

However, in general, we have

$$\delta\mathscr{L} \equiv \mathscr{L}'(x) - \mathscr{L}(x) = \frac{\partial\mathscr{L}}{\partial\phi}\delta\phi + \pi^\mu\delta\partial_\mu\phi,$$

so that to the lowest order in $\varepsilon$ we may write

$$4\varepsilon\mathscr{L}(x) + \varepsilon x\,.\,\partial\mathscr{L}(x) + O(\varepsilon^2) = \partial_\mu(\pi^\mu\delta\phi) = \partial_\mu\{\pi^\mu\varepsilon(d + x\,.\,\partial)\phi\} + O(\varepsilon^2),$$

or

$$\sum_\varphi \partial_\mu\{\pi^\mu\phi d + x_\nu\tilde{T}^{\mu\nu}\} = 0. \tag{3.18}$$

We can interpret eqn (3.18) as a current-conservation equation associated with the scale-invariance of the Lagrangian density, so that the dilation current $\mathscr{D}_\mu$ may be taken to be

$$\mathscr{D}_\mu = \pi_\mu\phi d + x^\nu\tilde{T}_{\mu\nu}.$$

For free or unrenormalized scalar fields ($d_\varphi = 1$) we have

$$\begin{aligned}\mathscr{D}_\mu &= \phi\partial_\mu\phi + x^\nu T_{\mu\nu}\\ &= x^\nu\{T_{\mu\nu} + \tfrac{1}{6}(g_{\mu\nu}\Box - \partial_\mu\partial_\nu)\phi^2\} - \partial_\nu(x_\mu\partial^\nu - x^\nu\partial_\mu)\phi^2.\\ &= x^\nu\theta_{\mu\nu} + \partial^\nu A_{\mu\nu},\end{aligned}$$

where $A_{\mu\nu}$ is an antisymmetric tensor. The divergence term in the above equation does not contribute to either $\partial^{\mu}\mathscr{D}_{\mu}$ or $D$—the two objects of measurable interest. Thus the second term, called a superpotential, can be ignored for all practical purposes. It can be shown (Jackiw 1971a) further that higher-spin fields do not contribute to $x^{\nu}(T_{\mu\nu}-\theta_{\mu\nu})$ either. Thus we arrive at the important relation mentioned earlier, namely,

$$\mathscr{D}_{\mu} \doteq x^{\nu}\theta_{\mu\nu}. \tag{3.19}$$

We can readily verify (see Exercise 3.1) eqn (3.19) for the theory defined by eqn (3.11).

*Canonical versus anomalous dimensions*

The scale dimension $d$ defines the representation of the group of scale transformations. It is, in principle, a continuous variable. In this sense it is different from the quantum numbers $j$, $m$ referring to a representation of the rotation group. The canonical dimension of a free or unrenormalized field is, of course, a fixed number. As discussed earlier, however, the occurrence of zero or infinite renormalization constants destroys equal-time commutation rules for the renormalized fields $\phi_{\mathrm{R}}$. Under these circumstances, there no longer exists any compelling reason for canonical scale dimensions to hold for renormalized fields. When the dimension of a renormalized field changes from its canonical value as a function of the strengths of interaction, it is termed an anomalous dimension. The hypothesis of anomalous dimensions, due to Wilson, states that dimensions of renormalized fields in a scale invariant theory with interactions *are* anomalous. For some very special fields, the anomaly may vanish because of certain fundamental symmetry constraints (e.g. for the stress–energy tensor $\theta_{\mu\nu}$ and for the $\mathrm{SU}(3)\times\mathrm{SU}(3)$ currents $J^{i}_{\mu}$ and $J^{i}_{\mu}$ and $J^{i}_{\mu 5}$—see Chapter 4, but this is not true in general. Wilson's hypothesis was based on the following considerations of renormalizable theories:

1. The Thirring model explicitly demonstrates this phenomenon (Wilson 1970b). This field-theoretical model involves a fermion field $\psi$ in one space and one time dimension only. It has a Fermi-type interaction:

$$\mathscr{L}_{1} = \lambda J^{\mu}J_{\mu}, \qquad \mu = 1, 2.$$

It is exactly solvable. By explicitly considering the two-point Green's function, we find

$$d_{\psi} = \frac{1}{2}+\frac{\lambda^{2}/4\pi^{2}}{1-(\lambda^{2}/4\pi^{2})}. \tag{3.20a}$$

2. A lowest-order perturbative approach to a theory of massless bosons with a $\lambda\phi^4$ interaction implies (Wilson 1970c) a departure from canonical dimensions, namely,

$$d_{\varphi^4} = 4+(\lambda g/\pi^2)+O(\lambda^2). \tag{3.20b}$$

3. Wilson has considered a non-perturbative approach to a theory of massless bosons with a $\lambda\phi^6$ interaction in a space of (2, 1) dimensions. Using his recently developed cut-off-based method, which has been used successfully in statistical mechanics, he has considered the renormalization of the two-point Green's function and found that $d_\varphi$ is anomalous (Wilson 1972).

On the other hand, Brandt and Preparata have hypothesized that, in the real world, renormalized fields retain their canonical dimensions. This hypothesis has been supported by several objections (Jackiw 1971b) raised against the anomalous-dimension idea. In cases (1) and (3) we are not dealing with a realistic four-dimensional space, and these calculations, especially the derivation of eqn (3.20a), may depend sensitively on the nature of the phase space. In obtaining eqn (3.20b) from a scale-symmetric Lagrangian density, use has been made of perturbation theory in the lowest order, where anomalies (Coleman 1971) force a departure from canonical dimensions. However, it may be a mistake to take perturbation theory seriously in this context, since in higher orders the same anomalies induce logarithmic violations of scale invariance itself. The most telling argument so far in favour of the Brandt–Preparata postulate is that, as discussed in the next chapter, the experimental data on deep inelastic eN scattering, obtained at SLAC, indicate canonical or near-canonical dimensions, at least for certain sets of fields.

If the scale dimension of a field does change because of interactions, from the Kallen–Lehmann representation (Bjorken and Drell 1965) we can obtain a lower bound on that dimension, namely, $d_\varphi \geqslant 1$. This can be seen in the following way. For renormalized fields associated with a particle of mass $\mu$, the said representation can be written as

$$\int \mathrm{d}^4x\, \mathrm{e}^{\mathrm{i}q.x}\langle 0|T\phi_{\mathrm{R}}^*(x)\phi_{\mathrm{R}}(0)|0\rangle = \frac{1}{q^2-\mu^2+\mathrm{i}\varepsilon}+\int_{4\mu^2}^{\infty} \mathrm{d}M^2\, \frac{\rho(M^2)}{q^2-M^2+\mathrm{i}\varepsilon}, \tag{3.21a}$$

where $\rho(M^2)$ is the positive definite spectral function

$$\rho(M^2) = \theta(M^2-4\mu^2)(2\pi)^3 \sum_n |\langle 0|\phi|n\rangle|^2. \tag{3.21b}$$

Eqns (3.21a) and (3.21b) imply that

$$\lim_{q\to\infty} \int \mathrm{d}^4x\, \mathrm{e}^{\mathrm{i}q.x}\langle 0|T\phi_{\mathrm{R}}^*(x)\phi_{\mathrm{R}}(0)|0\rangle \propto \frac{1}{q^\alpha}, \tag{3.22}$$

where $\alpha \leqslant 2$; the situation $\alpha < 2$ arises when the integral $\int_{4\mu^2}^{\infty} dM^2 \rho(M^2)$ diverges. By using a scale transformation in the variable of integration in eqn (3.22) and employing eqn (3.2), we easily obtain that

$$d_\varphi \geqslant 1. \tag{3.23}$$

Eqn (3.23) is, of course, not true for a $c$-number which has a vanishing scale dimension.

## Broken scale invariance

*Compulsion and consequences of scale-breaking*

So far, we have confined ourselves to a scale-invariant world. However, scale invariance cannot be an exact symmetry of the real world. This will be clear from the ensuing consideration. Let us return to eqn (3.6). On exponentiating this relation, we obtain

$$e^{i\varepsilon D} P^2 e^{-i\varepsilon D} = e^{2\varepsilon} P^2. \tag{3.24}$$

Take $|p\rangle$ to be a single-particle state of spin-$\frac{1}{2}$ and four-momentum $p$, so that

$$P^2|p\rangle = p^2|p\rangle.$$

Now, because of eqn (3.24), we may write

$$P^2 e^{-i\varepsilon D}|p\rangle = e^{2\varepsilon} p^2 e^{-i\varepsilon D}|p\rangle. \tag{3.25}$$

In other words, $\exp(-i\varepsilon D)|p\rangle$ is an eigenstate of $P^2$ with eigenvalue $\exp(2\varepsilon)p^2$. Moreover, if $a^\dagger(p)$ is the creation operator for the particle in question, we have

$$e^{-i\varepsilon D}|p\rangle = e^{-i\varepsilon D} a^\dagger(p)|0\rangle = e^{\frac{3}{2}\varepsilon} a^\dagger(e^\varepsilon p)\, e^{-i\varepsilon D}|0\rangle.$$

Hence, *if the vacuum is unique under scale transformations*, i.e. $\exp(-i\varepsilon D|0\rangle = |0\rangle$, then the operator $\exp(-i\varepsilon D)$ does not alter the nature of the particle in $|p\rangle$ but simply changes its momentum from $p$ to $(\exp\varepsilon)p$. Then, by virtue of eqn (3.25), all particles are massless or their mass spectra are continuous. Neither of these situations is realized in nature. Therefore, given a unique vacuum, scale symmetry has to be a broken symmetry in the real world. Hence the commutators of eqns (3.5)–(3.7) must be modified. However, the symmetry-breaking may be assumed to be gentle enough to preserve the algebraic field transformation of eqn (3.10). This then serves as a relation by which the dilation generator $D$ enters the theory. The only difference is that in a broken symmetry the generator $D$ becomes time-dependent. Thus the algebraic relations, which are maintained by the symmetry-breaking, have to be interpreted as equal-time commutators. Hence eqn (3.10) can be generalized to

$$[D(x^0), \phi(x)] = -i(d + x\,.\,\partial)\phi(x). \tag{3.26}$$

In order to keep track of the time-dependence in $D$, we shall henceforth write $D(x_0)$ instead of only $D$.

Let us now consider the modification of relations such as eqns (3.5)–(3.7) in a scale-broken theory. First introduce the derivative $\hat{\partial}_\mu$ which applies only with respect to the explicit dependence on $x$. Thus, with respect to any function $A(x)$, we may write

$$\hat{\partial}_\mu A(x) \equiv \partial_\mu A(x) - \mathrm{i}[P_\mu, A(x)]. \tag{3.27}$$

By definition, a local field $\phi(x)$ has no such dependence, so that $\hat{\partial}_\mu \phi(x) = 0$, and we obtain

$$\partial_\mu \phi(x) = \mathrm{i}[P_\mu, \phi(x)]. \tag{3.28}$$

The application of $\hat{\partial}_\mu$ to both sides of eqn (3.26) leads to

$$\mathrm{i}[\hat{\partial}_\mu D(x^0), \phi(x)] = \partial_\mu \phi(x) \tag{3.29}$$

for all local fields $\phi(x)$. Eqns (3.28) and (3.29) imply that $\hat{\partial}_\mu D(x_0) = P_\mu$, and comparing with eqn (3.27) we finally obtain

$$\mathrm{i}[D(x^0), P_0] = P_0 - \frac{\mathrm{d}}{\mathrm{d}x^0} D(x^0) = P_0 - \int \mathrm{d}^3 x\, \partial \,.\, \mathscr{D}$$

and

$$\mathrm{i}[D(x^0), P_i] = P_i.$$

These four relations may be written together as

$$\mathrm{i}[D(x^0), P_\mu] = P_\mu - g_{\mu 0} \int \mathrm{d}^3 x \partial \,.\, \mathscr{D}. \tag{3.30}$$

Exact scale invariance means a conserved dilation current, i.e. $\partial \,.\, \mathscr{D} = 0$, so that we recover eqn (3.5). Further, in a scale-broken world, we similarly (see Exercise 3.2) have

$$[D(x^0), M_{\mu\nu}] = \int \mathrm{d}^3 x (g_{\mu 0} x_\nu - g_{\nu 0} x_\mu) \partial \,.\, \mathscr{D}. \tag{3.31}$$

Thus in a scale-broken world $D$ is no more a Lorentz scalar.

We have related the breaking of scale symmetry to the non-vanishing divergence of the dilation current $\partial \,.\, \mathscr{D}$ as well as to the existence of non-zero masses. The connection between the latter two is seen by taking the said divergence between single-particle states of mass $m$ and performing a spatial integration. Discarding the surface term and using eqn (3.30), we may write

$$\begin{aligned}\int d^3x\, \langle p'|\partial\, .\, \mathscr{D}|p\rangle &= \frac{d}{dx^0}\langle p'|D(x^0)|p\rangle \\ &= \langle p'|H|p\rangle - i\langle p'|[D(x^0), H]|p\rangle \\ &= p_0\delta^{(3)}(\mathbf{p}'-\mathbf{p}) - i(p_0 - p'_0)\langle p'|D|p\rangle. \end{aligned} \tag{3.32}$$

We now use the space part of eqn (3.6), namely,

$$-2i\mathbf{P}^2 = [D, \mathbf{P}^2],$$

between the said single-particle states to obtain

$$\begin{aligned}\frac{\mathbf{p}^2}{p^0}\delta^{(3)}(\mathbf{p}-\mathbf{p}') &= \frac{i}{2p^0}\langle p'|[D, \mathbf{P}^2]|p\rangle \\ &= \frac{i(\mathbf{p}^2 - \mathbf{p}'^2)}{2p^0}\langle p'|D|p\rangle \\ &= i(p_0 - p'_0)\langle p'|D|p\rangle. \end{aligned} \tag{3.33}$$

The substitution of eqn (3.33) into eqn (3.32) leads, with $p^2 = m^2$, to

$$\int d^3x\langle p'|\partial\, .\, \mathscr{D}|p\rangle = \left(\frac{1}{2\pi}\right)^3 \frac{\delta^{(3)}(\mathbf{p}-\mathbf{p}')}{p^0} m^2. \tag{3.34}$$

In eqn (3.34) the connection between mass and the divergence of the dilation current is quite manifest. Finally, let us remark that the trace of the stress–energy tensor $\theta^\mu_\mu$ is also a measure of scale-breaking, i.e. that the said tensor is traceless in the limit of exact scale invariance. This is a consequence of eqn (3.20) from which it follows that

$$\partial\, .\, \mathscr{D} = \theta^\mu_\mu,$$

or

$$\frac{dD}{dx_0} = \int d^3x\, \theta^\mu_\mu. \tag{3.35}$$

*Small scale-breaking and residual scale invariance*

For broken scale symmetry to be a useful symmetry, it is necessary (Mack 1967) that nature very 'nearly' respect scale invariance. In other words, it makes sense to decompose the strong-interaction Hamiltonian density into two parts, i.e. $\theta_{00} = \theta^{(0)}_{00} + \dots$ . Here $\theta^{(0)}_{00}$ is 'large' in some sense and is scale invariant. The additional piece is 'small' and breaks scale invariance. Thus, in the limit of exact symmetry, the scale-breaking part 'smoothly' goes to zero. The scale invariance of $\theta^{(0)}_{00}$ is formally expressed by the relation

$$[D, \theta^{(0)}_{00}] = -i(4 + x\, .\, \partial)\theta^{(0)}_{00}. \tag{3.36}$$

The difference $\theta_{00}-\theta_{00}^{(0)}$ is responsible for the masses of particles. This follows since in an exactly scale-invariant world all masses vanish. For other operators we may write

$$O(x) = O^{(0)}(x)+ \dots .$$

The difference $O(x)-O^{(0)}(x)$ may be parametrized in terms of the same coupling constants that appear in $\theta_{00}(x)-\theta_{00}^{(0)}(x)$. These go to zero in the symmetry limit, so that all operators $O(x)$ must tend to $O^{(0)}(x)$. In that limit then we are left with the residual scale invariant theory, which is called the *skeleton theory*. A general $O(x)$ need not have any definite dimension, but $O^{(0)}(x)$ must have one if dimensions are defined in the skeleton theory. The scale dimension of $O^{(0)}(x)$ is sometimes called the 'asymptotic' dimension of $O(x)$.

We now consider the above ideas more quantitatively. Let us write

$$\theta_{00} = \theta_{00}^{(0)}+\sum_l \lambda_l w_l, \tag{3.37}$$

where $w_l$ are local Lorentz scalar operators with corresponding dimensions $d_l$. We thus have the equal-time commutator

$$[D(x_0), w_l(x)] = -\mathrm{i}(d_l+x\,.\,\partial)w_l(x). \tag{3.38}$$

Therefore, using eqns (3.35), (3.36), and (3.38), we obtain

$$\begin{aligned}\int \mathrm{d}^3x\,\theta^\mu_\mu &= \int \mathrm{d}^3x\,\partial\,.\,\mathscr{D}\\ &= \frac{\mathrm{d}D}{\mathrm{d}x^0}\\ &= P_0-\mathrm{i}[D, P_0]\\ &= \int \mathrm{d}^3x\left(\theta_{00}^{(0)}+\sum_l \lambda_l w_l\right)-\\ &\quad -\int \mathrm{d}^3x\left\{(4+x\,.\,\partial)\theta_{00}^{(0)}+\sum_l (d_l+x\,.\,\partial)\lambda_l w_l\right\}.\end{aligned}$$

After partial integration this leads to

$$\int \mathrm{d}^3x\,\theta^\mu_\mu = \int \mathrm{d}^3x\sum_l (4-d_l)\lambda_l w_l. \tag{3.39}$$

We now use the theorem (which is in fact a corollary of the Federbush–Johnson theorem) that, for a local Lorentz scalar or pseudoscalar field operator $S(x)$, $\int S(x)\,\mathrm{d}^3x = 0$ implies that $S(x) \equiv 0$. From eqn (3.39) we then have

$$\theta^\mu_\mu = \sum_l (4-d_l)\lambda_l w_l. \tag{3.40}$$

If we now write

$$\theta_{\mu\nu} = \theta^{(0)}_{\mu\nu} + a g_{\mu\nu} + b g_{\mu 0} g_{\nu 0}$$

and use eqn (3.40) along with the relations $\theta^{\mu(0)}_{\mu} = 0$ and

$$\theta_{00} = \theta^{(0)}_{00} + \sum_{l} \lambda_l w_l,$$

we obtain

$$\theta_{\mu\nu} = \theta^{(0)}_{\mu\nu} + g_{\mu\nu} \sum_{l} \lambda_l w_l - \tfrac{1}{3}(g_{\mu\nu} - g_{\mu 0} g_{\nu 0}) \sum_{l} d_l \lambda_l w_l. \tag{3.41}$$

It may be seen from eqn (3.41) that $\theta^{(0)}_{\mu\nu}$ is not Lorentz covariant. This is to be expected from the fact that $D$ is not Lorentz invariant in a scale-broken world.

*Goldstone's theorem for scale invariance*

The theorem of Goldstone for dilation invariance may be stated as follows (Zumino 1970): in the limit of exact scale symmetry for the Lagrangian density either all masses are zero or there exists a massless scalar boson. For a simple but elegant illustration of this theorem let us consider the matrix element of $\theta_{\mu\nu}$ between single-particle states of momenta $p_1$ and $p_2$, spin-0, and mass $m$. Define $P \equiv \frac{1}{2}(p_1 + p_2)$ and $K = p_2 - p_1$. Then, using the divergenceless property of the stress–energy tensor, we can write the said matrix element in terms of two form factors as follows:

$$\langle p_2 | \theta_{\mu\nu} | p_1 \rangle = \left(\frac{1}{2\pi}\right)^3 \frac{1}{\sqrt{(4 p_{1,0} p_{2,0})}} \{2 P_\mu P_\nu F(K^2) + (K_\mu K_\nu - g_{\mu\nu} K^2) G(K^2)\}. \tag{3.42}$$

In eqn (3.42) the normalization $\langle p | \theta_{00} | p \rangle = (1/2\pi)^3 m$ is ensured by requiring $F(0) = 1$. This equation now implies the following expression for the matrix element of the trace of the stress–energy tensor:

$$(2\pi)^3 \sqrt{(4 p_{1,0} p_{2,0})} \langle p_2 | \theta^\mu_\mu | p_1 \rangle = (2m^2 - \tfrac{1}{2}K^2) F(K^2) - 3K^2 G(K^2). \tag{3.43}$$

In the limit of exact scale invariance, $\theta^\mu_\mu = 0$ and eqn (3.43) admits of two possibilities:

1. If $G(0)$ is finite,

$$\langle p | \theta^\mu_\mu | p \rangle = m^2 / p_0 = 0, \quad \text{or} \quad m = 0$$

for the scale-symmetric situation. This is the usual condition of vanishing masses in a dilation-invariant world.

2. Suppose $G(K^2)$ has a pole at $K^2 = 0$, i.e. $G(K^2) \simeq g/K^2$ for $K^2$ near zero. In this case, eqn (3.43) leads to

$$0 = \langle p | \theta^\mu_\mu | p \rangle = m^2 - \tfrac{3}{2} g / p_0 .$$

In other words $m^2 = \frac{3}{2}g$, so that $m$ is non-zero. In this case, however, because of the $1/K^2$ factor in $G(K^2)$ we must have a massless scalar boson (called the dilaton) which couples to the energy–momentum tensor as illustrated in Fig. 33.

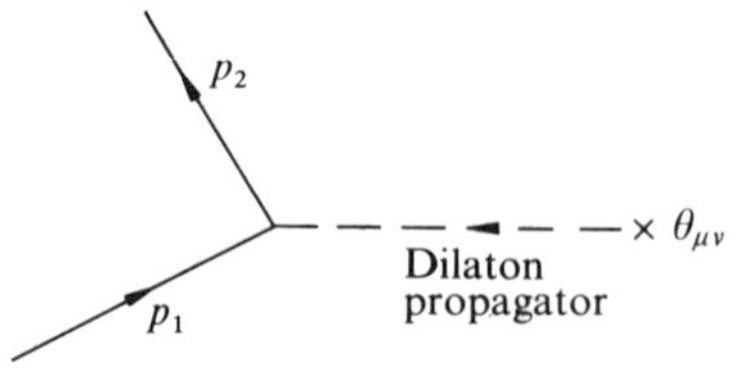


FIG. 33. Coupling of the massless scalar boson (dilaton) to the energy–momentum tensor.

The situation in the second case discussed above is exactly parallel to that in Goldstone-realized chiral SU(2) symmetry. In the latter case, we have a massless pion and the Goldberger–Treiman result (Adler and Dashen 1968). Here we have a massless scalar boson and the relation $m^2 = \frac{3}{2}g$. The matrix element of the dilation charge $D$ between the vacuum and the dilaton is non-zero, i.e. the vacuum is broken by $D$. Let us repeat that, if scale symmetry is Goldstone-realized, then the vacuum is not scale invariant in the exact symmetry limit of the Lagrangian density. Starting from a given vacuum, new vacua can be obtained by adding massless dilatons. If a massless scalar meson is involved in the limit of exact scale invariance, then we may wonder if it is identifiable in the real world as a massive scalar meson coupled to the stress–energy tensor. The scalar meson $\varepsilon(700)$ could be a possible candidate.

## Exercises

3.1. Verify the connection between the dilation current and the stress–energy tensor (eqn (3.19)) for the theory defined by the Lagrangian density of eqn (3.11).

3.2. Show that in a scale-broken world the commutator between the dilation charge and the generator of homogeneous Lorentz rotation is given by eqn (3.31).

# 4

# SHORT-DISTANCE SCALE INVARIANCE

## Essence

IN field theory the phrase 'short distances' is used in connection with ordinary or ordered products of local operators. It refers to the differences between the coordinates of those operators being small. Such small distances turn out to be relevant in amplitudes or cross-sections when their conjugate momenta become very large. This situation pertains to reactions where both energies and certain effective masses become large, as can be attained in lepton–hadron processes. Purely hadronic reactions, which cover regions with high energies and fixed masses, cannot probe exclusively short distances. For the latter reactions, scale invariance has no direct consequences. Indeed, a naive application of that idea would suggest total hadronic cross-sections falling off as $E_{\text{CM}}^{-2}$ at high energies, in direct contradiction with observation. It is at short distances that the idea of leading scale invariance makes physical sense. Let us discuss the essence of this idea. This is contained in the following statement (Wilson 1969, 1970*a*): the leading terms in products of operators in the asymptotic limit of small distances are the same as those given by the skeleton theory. This may be made concrete by taking the simplest case of a product of two operators $O(x)$ and $\bar{O}(y)$. Then we may write

$$\lim_{x-y\to 0} O(x)\bar{O}(y) = O^{(0)}(x)\bar{O}^{(0)}(y) + \text{non-leading terms.} \tag{4.1}$$

A product of operators at the same point, e.g. $O(x)\bar{O}(x)$, is not defined because of its highly divergent nature; the left-hand side of eqn (4.1) develops a singularity as $x$ tends to $y$. However, the degree of this singularity is being claimed to be the same as that in the dilation-invariant skeleton theory. This enables us to obtain a quantitative knowledge of the power of the said singularity by use of dimensional analysis. The strength of the hypothesis stated earlier will become clear later when we come to operator product expansions. But first let us provide some justifications for the idea itself, given a world of broken scale invariance with visibly non-zero masses. This we shall do by recourse to free field theory as well as to the renormalization group equation.

*Free field theory arguments*

We start with the scale invariant theory of free massless bosons. The equation of motion

$$\Box\phi = 0 \tag{4.2}$$

displays manifest scale invariance. However, under the transformations of eqns (3.1) and (3.2), the equal-time commutator

$$[\phi(x), \dot{\phi}(x^0, \mathbf{y})] = \mathrm{i}\delta^{(3)}(\mathbf{x}-\mathbf{y}) \tag{4.3}$$

remains unchanged and the Hamiltonian

$$H = \tfrac{1}{2}\int \mathrm{d}^3x\{\dot{\phi}^2+(\nabla\phi)^2\} \tag{4.4}$$

transforms properly (i.e. $U(D)H(x^0)U(D)^{-1} = \mathrm{e}^{\varepsilon}H(\mathrm{e}^{\varepsilon}x^0)$) only when the field has the canonical dimension $d_{\varphi} = 1$. The scale invariance of the two-point function

$$D_{\mathrm{F}}(x) = \langle 0|T\phi^*(x)\phi(0)|0\rangle. \tag{4.5}$$

may also be investigated. Taking the vacuum to be invariant under dilations, we can write

$$\begin{aligned} D_{\mathrm{F}}(x) &= \langle 0|TU(D)\phi^*(x)U(D)^{-1}U(D)\phi(0)U(D)^{-1}|0\rangle \\ &= \mathrm{e}^{2\varepsilon}\langle 0|T\phi^*(\mathrm{e}^{\varepsilon}x)\phi(0)|0\rangle, \end{aligned} \tag{4.6}$$

or

$$D_{\mathrm{F}}(x) = \mathrm{e}^{2\varepsilon}D_{\mathrm{F}}(\mathrm{e}^{\varepsilon}x). \tag{4.7}$$

Eqn (4.7) implies that $D_{\mathrm{F}}(x)$ scales as $x^{-2}$ times a dimensionless constant. In momentum space we may write

$$\begin{aligned} D_{\mathrm{F}}(p) &= \int \mathrm{d}^4x\, \mathrm{e}^{\mathrm{i}p.x}D_{\mathrm{F}}(x) \\ &= \int \mathrm{d}^4x\, \mathrm{e}^{\mathrm{i}p.x}\, \mathrm{e}^{2\varepsilon}D_{\mathrm{F}}(e^{\varepsilon}x), \end{aligned} \tag{4.8}$$

or

$$\begin{aligned} D_{\mathrm{F}}(p) &= \mathrm{e}^{-2\varepsilon}\int \mathrm{d}^4y\, D_{\mathrm{F}}(y)\exp(\mathrm{i}\,\mathrm{e}^{-\varepsilon}py) \\ &= \mathrm{e}^{-2\varepsilon}D_{\mathrm{F}}(\mathrm{e}^{-\varepsilon}p). \end{aligned} \tag{4.9}$$

In other words $D_{\mathrm{F}}(p)$ scales as $p^{-2}$ times a dimensionless constant. In the real world, because of a mass term in the Lagrangian density, the equation of motion $(\Box+m^2)\phi = 0$ breaks scale invariance. In the propagator $D_{\mathrm{F}}(p)$, $p^{-2}$ is changed to $(p^2-m^2)^{-1}$. Hence, for values of $p^2$ of the order of $m^2$, $D_{\mathrm{F}}(p)$ is very different from that in the massless case. However, for large values of $p^2$ that are much greater than $m^2$ (i.e. for highly virtual masses), the leading term in the former coincides with the latter. In coordinate space, the $x^{-2}$ in $D_F(x)$ gets changed to

$$x^{-2}+\frac{m^2}{2\pi}\ln(m^2x^2) \tag{4.10}$$

on introducing a mass. However, leading scale invariance is still maintained in the limit of vanishing $x$. In the case of spin-$\frac{1}{2}$ fields the form of the propagator in configuration space is

$$S_F(x) = -i\partial\!\!\!/ D_F(x) \tag{4.11}$$

for massless fields and

$$S_F(x) = -(i\partial\!\!\!/+m)D_F(x) \tag{4.12}$$

in the massive case. The former is proportional to $x\!\!\!/ x^{-4}$ and the latter to

$$\left(x\!\!\!/+\frac{im}{2}x^2\right)x^{-4}-\frac{m^3}{4\pi i}\ln(m^2x^2)-\frac{m^2}{\pi}x\!\!\!/ x^{-2}. \tag{4.13}$$

Once again, leading scale invariance follows when $x$ approaches zero.

*Renormalization group arguments*

The validity of asymptotic scale invariance in interacting field theories of massive particles can be demonstrated by considering the renormalization group equation and its solution (Wilson 1971; Coleman 1971; Callan 1972a; Mack 1972; Schroer 1972). The full argument is somewhat involved and we follow here a simplified heuristic treatment of the subject, after Coleman. Consider the theory of a self-interacting massive scalar field $\phi$, as given by the Lagrangian density

$$\mathscr{L} = \tfrac{1}{2}\partial\phi\,.\,\partial\phi-\frac{\lambda_0}{4!}\phi^4-\tfrac{1}{2}\mu_0^2\phi^2. \tag{4.14}$$

(This specific instance is chosen for simplicity and definiteness; the related discussion can be generalized to more complicated cases without much difficulty.) We take the sum of all connected Feynman diagrams with external lines carrying momenta $p_1, \ldots, p_n$. This may be denoted by $T^{(n)}(p_1, \ldots, p_n)$, where

$$(2\pi)^4\delta^{(4)}\left(\sum_{i=1}^{n} p_i\right)T^{(n)}(p_1, \ldots, p_n)$$
$$= \int d^4x_1 \ldots \int d^4x_n \exp\left(i\sum_{i=1}^{n} p_i\,.\,x_i\right)\langle 0|T\phi(x_1)\ldots\phi(x_n)|0\rangle. \tag{4.15}$$

From an ordinary dimensional analysis of eqn (4.15), we can see that $T^{(n)}$ is $\mu^{-3n+4}$ times a dimensionless function, $\mu$ being the renormalized mass. Let us remove from $T^{(n)}$ the entire set of diagrams which can be separated into two disconnected pieces by cutting a single line. We call the remainder $\tilde{T}^{(n)}$, and divide it by all the external propagators. These operations leave us with the one-particle irreducible renormalized Green's function $\Gamma^{(n)}(p_1, \ldots, p_n)$ with $n$ external lines, where

$$\Gamma^{(n)}(p_1, \ldots, p_n) = \frac{\tilde{T}^{(n)}(p_1, \ldots, p_n)}{T^{(2)}(p_1) \ldots T^{(2)}(p_n)}. \tag{4.16}$$

Once again, ordinary dimensional analysis—when applied to eqn (4.16)—tells us that $\Gamma^{(n)}$ is $\mu^{-n+4}$ times a dimensionless function. More definitively, we may write

$$\Gamma^{(n)} = s^{(4-n)/2} F_n\left(\frac{s}{\mu^2}, \lambda, \frac{p_i \cdot p_j}{s}\right), \tag{4.17}$$

where $s \equiv \sum_{i=1}^{n} p_i^2$, $\lambda$ is the renormalized (dimensionless) coupling constant and $F_n$ is an unknown function. Corresponding to this renormalized $\Gamma^{(n)}$ there is an unrenormalized one-particle irreducible Green's function $\Gamma_u^{(n)}$ with similar dependence on the momenta and on the bare parameters. The two may be related by the standard power-counting arguments of renormalization theory (Coleman 1971) in the following way:

$$\Gamma^{(n)}(p_1, \ldots, p_n) = (Z_3)^{n/2} \Gamma_u^{(n)}(p_1, \ldots, p_n). \tag{4.18}$$

In eqn (4.18) $Z_3$ is the wavefunction renormalization constant which is dimensionless but depends on the regulator cutoff $\Lambda$ (necessary for handling infinities in the theory) to the bare mass $\mu_0$.

We now take the momentum-space equivalent of the short-distance limit. This entails going to the deep Euclidean (DE) region defined by the limit $s \to \infty$ with $(p_i \cdot p_j)/s$ fixed. It is a maximally unphysical domain, since here all the external lines are far off the mass shell. This latter fact suggests the emergence of scale invariance via the disappearance of any dependence of the Green's function on the meson mass in this limit. However, the renormalization of the field and of the coupling constant are defined on the mass shell, and some dependence on the mass must persist because of them. This calls for a certain amount of delicacy in handling the renormalized Green's function. The safest starting point to introduce the deep Euclidean mass-independence suggested above is therefore at the level of unrenormalized quantities. Thus we may require

$$\frac{\partial}{\partial \mu_0} \lim_{\mathrm{DE}} \Gamma_u^{(n)}(p_1, \ldots, p_n) = 0. \tag{4.19}$$

Eqn (4.19) can be derived more formally from perturbative considerations (Coleman 1971), but the above line of reasoning suggests that its validity extends beyond perturbation theory. In any event, eqns (4.18) and (4.19) lead to the result

$$\left(\frac{\partial}{\partial\mu_0}-\frac{n}{2}\frac{\partial \ln Z_3}{\partial\mu_0}\right)\lim_{\text{DE}}\Gamma^{(n)}(p_1,\ldots,p_n)=0. \qquad (4.20)$$

On multiplying either side of eqn (4.20) by $Z\mu_0$ (where $Z$ is a multiplicative constant) and using the chain rule of partial differentiation, we obtain

$$\left\{\left(Z\mu_0\frac{\partial\mu}{\partial\mu_0}\right)\frac{\partial}{\partial\mu}+\left(Z\mu_0\frac{\partial\lambda}{\partial\mu_0}\right)\frac{\partial}{\partial\lambda}-\frac{n}{2}\left(Z\mu_0\frac{\partial \ln Z_3}{\partial\mu_0}\right)\right\}\lim_{\text{DE}}\Gamma^{(n)}(p_1,\ldots,p_n)=0. \qquad (4.21)$$

The constant $Z$ is now chosen to obey the relation

$$Z\mu_0\frac{\partial\mu}{\partial\mu_0}=\mu. \qquad (4.22)$$

We further define

$$\beta\equiv Z\mu_0\frac{\partial\lambda}{\partial\mu_0}, \qquad (4.23)$$

$$\gamma\equiv\tfrac{1}{2}Z\mu_0\frac{\partial \ln Z_3}{\partial\mu_0}. \qquad (4.24)$$

Being coefficients in a differential equation satisfied by the renormalized Green's function, these must be independent of the regulator cutoff $\Lambda$. Hence dimensional considerations alone require them to be functions of $\lambda$ only. Finally then, we obtain the equation obeyed by the renormalized one-particle irreducible Green's function in the deep Euclidean region in terms of renormalized quantities only. This special case of the more general Callan–Symanzik equation (valid in all regions) reads

$$\left\{\mu\frac{\partial}{\partial\mu}+\beta(\lambda)\frac{\partial}{\partial\lambda}-n\gamma(\lambda)\right\}\lim_{\text{DE}}\Gamma^{(n)}(p_1,\ldots,p_n)=0. \qquad (4.25)$$

Eqn (4.25) is a linear first-order differential equation of the hydrodynamic type. It is not hard to solve. First, introduce the variables $t=\frac{1}{2}\ln(s/\mu^2)$ and $\lambda'\equiv\lambda'(\lambda,t)$ satisfying the equation

$$-\frac{\partial\lambda'(\lambda,t)}{\partial t}=\beta(\lambda') \qquad (4.26)$$

as well as the boundary condition

$$\lambda'(\lambda,0)=\lambda. \qquad (4.27)$$

The solution to eqn (4.25) may then be given in the form of eqn (4.17) taken in the DE limit, where we would have

$$\lim_{\mathrm{DE}} F_n\left(\frac{s}{\mu^2}, \lambda, \frac{p_i \cdot p_j}{s}\right) = f_n\left(\lambda'(\lambda, t), \frac{p_i \cdot p_j}{s}\right) \exp\left[-n \int_0^t \mathrm{d}t' \, \gamma\{\lambda'(\lambda, t')\}\right], \tag{4.28}$$

$f_n$ being an arbitrary function. Eqn (4.25) is referred to as the renormalization group equation because of the identification of the related eqn (4.26) with one of the Lie differential equations associated with the renormalization group, originally devised by Gell-Mann and Low in connection with quantum electrodynamics but later extended to general renormalizable field theories (Bogoliubov and Shirkov 1959). Note that *a priori* the asymptotic form of the Green's function could depend in an arbitrary way on the dimensionless quantities $\lambda$ and $t = \frac{1}{2}\ln(s/\mu^2)$. But the solution to the renormalization group equation restricts the arbitrary dependence only to a certain function of those variables, namely $\lambda'$—as determined by eqns (4.26) and (4.27).

The information content of all these equations can be extracted more fruitfully with the aid of two very reasonable assumptions. The first one is that the functions $f_n$ are continuous in $\lambda'$, and likewise $\beta$ and $\gamma$ are continuous in $\lambda$. The second assumption concerns the existence and location of zeros in $\beta(\lambda)$ on the positive real axis of the $\lambda$-plane. The Gell-Mann–Low function $\beta(\lambda)$ is known to have a quadratic zero at the origin. The region near $\lambda = 0$ may be legitimately handled by perturbation theory, which gives (Coleman 1971)

$$\beta(\lambda) = \frac{3\lambda^2}{16\pi^2} + O(\lambda^3). \tag{4.29}$$

According to eqn (4.29), $\beta(\lambda)$ becomes positive as $\lambda$ starts increasing from zero. Now assume that $\beta(\lambda)$ has at least one more zero at a positive real value (say $\lambda_1$) of $\lambda$. For the sake of definiteness we take $\lambda = 0, \lambda_1$ to be the only two zeros of $\beta(\lambda)$ for positive real $\lambda$. The presence of additional zeros complicates the following argument but does not alter the nature of the conclusion qualitatively. If $\lambda'(t = 0)$ is between 0 and $\lambda_1$, this information may be used in the integration of eqns (4.15). We see then that $\lambda'$ is forced to be a monotonically increasing function of $t$ which approaches $\lambda_1$ from below at $t \to \infty$; e.g. for $\beta(\lambda') \simeq (\lambda_1 - \lambda')$, near $\lambda' = \lambda_1$, $\lambda' \sim \lambda_1 + C\,\mathrm{e}^{-|a|t}$, where $C$ is an arbitrary constant. Similarly, if $\lambda'(t = 0)$ is between $\lambda_1$ and infinity, $\lambda'$ is forced to decrease monotonically for larger and larger values of $t$, approaching $\lambda_1$ from above as $t \to \infty$ (Fig. 34). Either way, the limiting

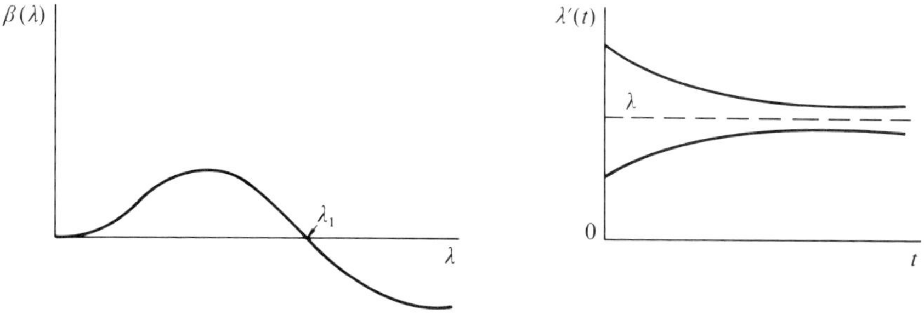


FIG. 34. The zero of the Gell-Mann–Low function as the asymptotic effective renormalized coupling constant.

value of $\lambda$ for large enough $t$ is seen to be $\lambda_1$. On using this information in eqns (4.17) and (4.28), we obtain

$$\lim_{\mathrm{DE}} \Gamma^{(n)} = s^{(4-n)/2} f_n\left(\lambda_1, \frac{p_i \cdot p_j}{s}\right)\left(\frac{s}{\mu^2}\right)^{-\frac{1}{2}n\gamma(\lambda_1)} K^n, \tag{4.30}$$

where

$$K = \exp \int_0^\infty \mathrm{d}t[\gamma(\lambda_1) - \gamma\{\lambda'(\lambda, t)\}]. \tag{4.31}$$

The functions $f_n$ are now finite in the limit of interest. The integral of eqn (4.31) is convergent if $\lambda_1$ is a first-order zero. This follows because in that case $\gamma(\lambda_1) - \gamma(\lambda')$ can be approximated by $(\lambda_1 - \lambda'_1)\gamma'(\lambda_1)$ at large $t$, and the factor in parenthesis is exponentially damped, as argued earlier. Under these circumstances $\lim_{\mathrm{DE}} \Gamma^{(n)}$ behaves as $s^{2-\frac{1}{2}n\{1+\gamma(\lambda_1)\}}$ times a scale function. (If $\lambda_1$ is a higher-order zero, we get the same power behaviour but with logarithmic functions of $s$ in addition, so that the limit becomes less interesting.) Comparing with eqn (4.17), we see that in the deep Euclidean region (or equivalently at short distances) the vertex function $\Gamma^{(n)}$ does exhibit scale invariance. The scale dimension of the renormalized field $\phi$ is seen to be $d_\phi = 1 + \gamma(\lambda_1)$, which has the canonical value only if $\gamma(\lambda_1)$ happens to vanish. This concludes, for the present, our discussions of scale invariance at short distances, based on the renormalization group equation. However, in Part III, in connection with light-cone physics, we shall return to the question of how canonical scale dimensions and indeed free field theory may be recovered at short distances for a different behaviour of $\beta(\lambda)$.

## Operator product expansions

The notion of scale invariance at short distances, as expounded above, is an elegant idea; but it is useful only if we can handle products of operators at those distances. This is best achieved through Wilson's operator product expansions (OPE), as explained in the following. We start with the statement of operator product expansions for the simplest case of the ordinary product of two operators. Given the local operators $A(x)$ and $B(y)$, it is possible to choose a set of local operators $\{O_n\}$ and a sequence of $c$-number functions $\{C_n(x-y)\}$ singular at $x = y$, such that we can write

$$A(x)B(y) \xrightarrow[x-y\to 0]{} \sum_n C_n(x-y)O_n(x) \tag{4.32}$$

in a weak sense (i.e. between physical states). To a finite order in $x-y$, only a finite number of fields $O_n$ are required. The validity of OPEs is supposed to rest on stronger grounds than asymptotic dilation symmetry. In field theory the validity of these expansions has been proved, with some minimal assumptions in addition to the standard axions (Zimmermann 1971). We give here a simplified version of that proof.

*Derivation of the OPE*

Let

$$x^\mu = z^\mu + \rho\eta^\mu, \tag{4.33a}$$

$$y^\mu = z^\mu - \rho\eta^\mu, \tag{4.33b}$$

with $\eta^2 \neq 0$ and $\rho > 0$. We introduce the matrix element

$$f_{\alpha\beta}(\rho, \eta) \equiv \langle\alpha|A(z+\rho\eta)B(z-\rho\eta)|\beta\rangle \tag{4.34}$$

between the states $\langle\alpha|$ and $|\beta\rangle$, and proceed to take the limit $\rho \to 0+$. The function $f_{\alpha\beta}(\rho, \eta)$ will be singular in this limit, but we assume that there exists some matrix element

$$f_0(\rho, \eta) \equiv \langle\alpha_0|A(z+\rho\eta)B(z-\rho\eta)|\beta_0\rangle \tag{4.35}$$

for which the ratio $f_{\alpha\beta}(\rho, \eta)/f_0(\rho, \eta)$ exists in the limit $\rho \to 0$. If there happen to be more than one $f_{\alpha\beta}(\rho, \eta)$ behaving equally strongly as $\rho$ tends to zero, this assumption requires the lack of any limiting oscillation in their ratios. We define the operator $O_0(z, \eta)$ as follows:

$$O_0(z, \eta) \equiv \lim_{\rho\to 0} \frac{A(z+\rho\eta)B(z-\rho\eta)}{f_0(\rho, \eta)}. \tag{4.36}$$

From eqn (4.36) we can immediately make two statements about this operator. (1) $O_0(z, \eta)$ cannot vanish identically since there exist at least two states $|\alpha_0\rangle, |\beta_0\rangle$ for which $\langle\alpha_0|O_0|\beta_0\rangle$ is unity. (2) $O_0(z, \eta)$ satisfies certain causal commutation relations, namely,

$$[O_0(z, \eta), A(u)] = 0$$

for $(z-u)^2 < 0$ and

$$[O_0(z,\eta), B(v)] = 0$$

for $(z-v)^2 < 0$. The second statement follows from the causal properties of $A$ and $B$. Its import is that $O_0$ is a causal operator for any permitted value of $\eta$; in other words it is a legitimate composite operator. Define $P_1(z, \rho\eta)$ through the relation

$$A(z+\rho\eta)B(z-\rho\eta) \equiv f_0(\rho\eta)O_0(z,\eta) + P_1(z,\rho\eta). \tag{4.37}$$

Eqn (4.37) has the interesting property that any singularity which develops in the first term of its right-hand side is contained solely in the function $f_0(\rho\eta)$ and leaves the operator $O_0(z,\eta)$ undisturbed. It further follows from this equation that

$$\lim_{\rho\to 0}\frac{P_1(z,\rho\eta)}{f_0(\rho\eta)} = \lim_{\rho\to 0}\frac{A(z+\rho\eta)B(z-\rho\eta)}{f_0(\rho\eta)} - O_0(z,\eta) = 0.$$

It is thus clear that the first term in the right-hand side of eqn (4.37) is more dominant in the limit concerned than the second term.

We may now repeat this procedure. First we define

$$O_1(z,\eta) \equiv \lim_{\rho\to 0}\frac{P_1(z,\rho\eta)}{f_1(\rho\eta)}. \tag{4.38}$$

In eqn (4.38), $f_1(\rho\eta)$ is defined with respect to $P_1$ in the same way that $f_0(\rho\eta)$ was defined *vis à vis* $A(z+\rho\eta)B(z-\rho\eta)$ in eqn (4.35). Then we may write

$$P_1 = f_1 O_1 + P_2,$$

where

$$\lim_{\rho\to 0}\frac{f_1}{f_0} = 0 = \lim_{\rho\to 0}\frac{P_2}{f_1}.$$

Hence we are led to the result

$$A(z+\rho\eta)B(z-\rho\eta) = f_0 O_0 + f_1 O_1 + P_2.$$

The iteration of the above procedure for $k$ times yields

$$A(z+\rho\eta)B(z-\rho\eta) = \sum_{i=0}^{k} f_i(\rho\eta)O_i(z,\eta) + P_{k+1}(z,\rho\eta), \tag{4.39}$$

where

$$\lim_{\rho\to 0}\frac{f_{i+1}}{f_i} = 0 = \lim_{\rho\to 0}\frac{P_{k+1}}{f_k}. \tag{4.40}$$

Eqn (4.39), given the conditions of eqn (4.40), is the desired OPE (cf. eqn (4.32)), which has a finite number of terms to any specified order in $\rho^{-1}$. The main assumption involved in the derivation is that at every step $k$ there is a matrix element of $P_k$ which behaves in the $\rho \to 0$ limit more strongly than, or at least as strongly as, any other matrix element of $P_k$. This can be weakened into a condition on the matrix elements of $A(x)B(y)$ (Wilson and Zimmermann 1972) but cannot be disposed of entirely.

*Discussions of the OPE*

We first give an example. Consider a free scalar field theory. Using Wick's theorem, we have the identity

$$:\phi^2(x)::\phi^2(y): = 2\{D(x-y)\}^2 I + 4D(x-y):\phi(x)\phi(y): + :\phi^2(x)\phi^2(y):, \quad (4.41)$$

where

$$D(x-y) = \langle 0|\phi(x)\phi(y)|0\rangle \propto (x-y)^{-2}.$$

Expanding the Wick products in Taylor series, we can write

$$:\phi(x)\phi(y): = :\phi^2(x): + (y-x)_\mu :\phi(x)\partial^\mu\phi(x): + \cdots,$$

etc. However, not all operators thus obtained are linearly independent. By a straightforward but tedious procedure we accomplish the reduction of the operators into a linearly independent set and finally obtain the required OPE.

Next, we can draw on a classical analogy. The operator product expansion is similar in spirit to the large distance expansion in classical potential theory. There we write the potential $V(\mathbf{r})$ as

$$V(\mathbf{r}) = \int \frac{d^3r'\rho(\mathbf{r}')}{|\mathbf{r}-\mathbf{r}'|} = \frac{Q}{r} + \frac{\mathbf{r}\,.\,\mathbf{d}}{r^3} + \cdots,$$

where $\rho(\mathbf{r})$ is the density, $Q$ the charge, and $\mathbf{d}$ the dipole moment of the source. The Coulomb potential dominates the expansion unless $Q$ happens to be zero, in which case the dipole potential dominates, and so on. A similar situation holds here. The major distinction is that, whereas terms such as $r^{-1}$, $r^{-3}$, etc. are non-singular as $r \to \infty$, our $c$-number functions $C_n(x-y)$ are highly singular as $x-y$ approaches zero.

The technique of using OPEs becomes a very potent one when combined with Wilson's proposal of scale invariance as an exact symmetry at short distances. Applying the unitary scale transformation operator $U(D)$ to both sides of eqn (4.32) and using the transformation properties of the operators, we obtain the nature of the short-distance singularity in $C_n(x)$, i.e.

$$C_n(x) \simeq Cx^{-d_A-d_B+d_n}, \quad (4.42)$$

as $x$ tends to zero, $C$ being an unknown constant. In eqn (4.42) the quantities $d_A$, $d_B$, and $d_n$ stand for the scale dimensions of $A(x)$, $B(x)$, and $O_n(x)$ respectively, as defined in the skeleton theory (these will be referred to as the 'asymptotic scale dimensions' of those operators). Hence the order of importance of any particular operator in the set $\{O_n\}$ is determined by the smallness of its dimension; the latter is what controls the power of the singularity in $C_n(x-y)$. Here we wish to emphasize the necessity for considering renormalized fields for $O_n$. Only then can the matrix elements $\langle\alpha|O_n|\beta\rangle$ be obtained without explicit renormalization constants $Z$. Thus the quantity $d_n$ in eqn (4.42) is the dimension of the renormalized field, and may be anomalous. In interacting field theories which are renormalizable the OPE together with the dimensional rule of eqn (4.42) can be derived from the renormalization group equation (Coleman 1971), but the $d$'s are anomalous in general.

The OPE can be generalized to the case of a larger number of operators. For example, in analogy with eqn (4.32) we may write

$$A(x)B(y)C(z) \underset{\substack{x-y\to 0,\\ x-z\to 0}}{\simeq} \sum_n f_n(x-y, x-z)O_n(z), \tag{4.43}$$

where the short-distance singularities of $f_n(x-y, x-z)$ can be determined by scale invariance up to arbitrary dependence of dimensionless variables such as $(x-z)^2/(x-y)^2$.

Finally, eqn (4.42) may not be completely consistent with an unbounded mass spectrum, and some refinement is called for. To illustrate the point in question take the matrix element of eqn (4.32) between the states $\langle\alpha|$ and $|\beta\rangle$:

$$\langle\alpha|A(x)B(0)|\beta\rangle = \sum_n C_n(x)\langle\alpha|O_n(0)|\beta\rangle. \tag{4.44}$$

Inserting a complete set of states, the left-hand side of eqn (4.44) may be written as

$$\sum_m \langle\alpha|A(x)|m\rangle\langle m|B(0)|\beta\rangle = \sum_m \exp\{\mathrm{i}(p_\alpha - p_m)\,.\,x\}\langle\alpha|A(0)|m\rangle\langle m|B(0)|\beta\rangle.$$

The above summation may not converge as the energy $E_m \to \infty$. The replacement of $x_0$ by $x_0 - \mathrm{i}\varepsilon$ provides the convergence factor $\exp(-\varepsilon E_m)$. Defining $2p = d_A + d_B - d_n$, we then obtain

$$C_n(x) = C(-x^2 + \mathrm{i}\varepsilon x_0)^{-p}. \tag{4.45}$$

Eqn (4.45) describes the form at short distances of the singular $c$-number function appearing in the expansion for the ordinary product of two operators. For the commutator of two hermitian operators $A$ and $B$, it follows from the Schwarz reflection principle that

$$[A(x), B(0)] \underset{x\to 0}{\simeq} \sum_n E_n(x)O_n(0), \tag{4.46}$$

where

$$\begin{aligned} E_n(x) &= C\{(-x^2+\mathrm{i}\varepsilon x^0)^{-p} - (-x^2 - \mathrm{i}\varepsilon x^0)^{-p}\}, \\ &= C_n(x) - C_n^*(x) \\ &= C_n(x_0 - \mathrm{i}\varepsilon, \mathbf{x}) - C_n(x_0 + \mathrm{i}\varepsilon, \mathbf{x}). \end{aligned} \tag{4.47}$$

The singularity structure of the function $E_n(x)$ is clear from eqn (4.47). It vanishes for $x^2 < 0$, as required by causality, and has a cut on the positive real $x^2$-axis. We can make a similar statement for the time-ordered product of $A$ and $B$. In that case, we obtain (see Exercise 4.1)

$$TA(x)B(0) \underset{x\to 0}{\simeq} \sum_n D_n(x)O_n(0), \tag{4.48a}$$

$$D_n(x) \underset{x\to 0}{\simeq} C\,(-x^2+\mathrm{i}\varepsilon)^{-p} \tag{4.48b}$$

with $p$ as defined earlier. Another interesting point is the connection between these OPEs and the equal-time commutator between $A$ and $B$. In the limit $x \to 0$, we may specialize further by taking $x_0$ to vanish faster than $x_i$. The existence of Schwinger terms at equal times—involving space derivatives of delta functions—is linked with the nature of the power singularity at short distances. In a more definitive manner, it may be shown (see Exercise 4.2) that

$$[A(0, \mathbf{x}), B(0)] = \sum_n \left( E_n^0 + \sum_{l=1} E_n^{i_1 \ldots i_l} \partial_{i_1 \ldots i_l} \right) \delta^{(3)}(\mathbf{x}) O_n(0), \tag{4.49}$$

where

$$E_n^0 = \int \mathrm{d}^3 z\, E_n(0, \mathbf{z}), \tag{4.50a}$$

$$E_n^{i_1 \ldots i_l} = \frac{(-1)^l}{l!} \int \mathrm{d}^3 z\, z^{i_1} \ldots z^{i_l} E_n(0, \mathbf{z}), \tag{4.50b}$$

and the $E_n(z)$ are as defined in eqn (4.47). It further follows from eqns (4.49) and (4.50) that $E_n^0$ has a zero, non-zero finite, or infinite value for $p$ less than, equal to, or greater than 1·5, respectively. Similar statements can be made on the other $E_n$s for higher values of $p$. So long as $p$ is finite, there will only be a finite number of Schwinger terms.

### The Wilson framework

Wilson introduced a set of hypotheses (Wilson 1969, 1970a) combining the ideas of scale invariance and OPEs at short distances. An important new element in his scheme, which has proved productive, is his enumeration of the basis set of operators $\{O_n(x)\}$. These are chosen to be those fields which can be abstracted from Wick products of elementary quark fields and which retain the Lorentz, internal, and discrete symmetry properties (though not necessarily the canonical scale dimensions) of the latter. In addition, a detailed investigation of renormalized perturbation theory led Wilson to the following results on the scale-breaking part $\mathscr{H}_{\mathrm{SB}}$ of the time–time component of the stress–energy tensor $\theta_{\mu\nu}$. (1) If the scale dimension $d_{\mathscr{H}_{\mathrm{SB}}}$ of the term in question exceeds 4, the scale-breaking interactions are non-renormalizable; to every order the leading ultraviolet singularity exhibits violations—which are power-wise or worse—of asymptotic scale invariance. (2) If $d_{\mathscr{H}_{SB}}$ equals 4, the scale-breaking forces are renormalizable and short-distance scale invariance holds up to logarithmic terms. (3) In case $d_{\mathscr{H}_{SB}}$ is less than 4, the scale breakers are super-renormalizable, and scale invariance becomes an exact symmetry at short distances. Wilson's proposal is that the part of the Hamiltonian density that violates dilation symmetry is of type (3). Indeed, according to Wilson, this part consists of operators which have the structure of mass terms (i.e., in quark language, the form $\bar{q}\mathscr{M}q$, where $\mathscr{M}$ is the mass matrix) which have the canonical scale dimension of 3. Hence, at least in the Brandt–Preparata hypothesis (see Chapter 3) of canonical scale dimensions for renormalized operators, the possibility of type (3) is realized. Those committed to the introduction of the gluon's field $B_\mu$ may also add an operator, abstracted from the gluon mass term $\frac{1}{2}\mu_g^2 B^2$, whose canonical scale dimension is 2.

*Composition of the Wilson framework*

We list below the ingredients constituting the Wilson framework which are relevant to our considerations:

(1) the scale invariant OPE existing at short distances for ordinary and ordered products (binary, trinary, etc.) and for commutators of local operators;

(2) the assumed form of the energy density as

$$\theta_{00} = \theta_{00}^{(0)} + \mathscr{H}_{SB},$$

where $d_{\mathscr{H}_{SB}} < 4$;

(3) the skeleton energy density $\theta_{00}^{(0)}$, invariant under both dilation and $SU(3)\times SU(3)$ symmetry, making the scale invariant OPE at short distances also invariant under the group $SU(3)\times SU(3)$;

(4) an enumeration of fields hermitian $O_n$, with scale dimensions less than or equal to 4, as follows,

| Operator | Quark structure | Nomenclature | Asymptotic scale dimensions |
|---|---|---|---|
| $I$ | | unit operator | 0 |
| $\theta_{\mu\nu}$ | $\mathrm{i}{:}\bar{q}(\gamma_\mu\overleftrightarrow{\partial}_\nu+\gamma_\nu\overleftrightarrow{\partial}_\mu)q{:}$ | stress–energy tensor | 4 |
| $J^i_\mu, J^i_{\mu5}$ | ${:}\bar{q}(\gamma_\mu, \gamma_\mu\gamma_5)\frac{\lambda^i}{2}q{:}$ | hadronic vector and axial vector currents (first-class) | 3 |

| Operator | Quark structure | Nomenclature | Asymptotic scale dimensions |
|---|---|---|---|
| $u^i, v^i$ | $:\bar{q}(1, \mathrm{i}\gamma_5)\frac{\lambda^i}{2}q:$ | scalar and pseudoscalar densities | 3? |
| $A^i_{\mu\nu}$ | $:\bar{q}\sigma_{\mu\nu}\frac{\lambda^i}{2}q:$ | antisymmetric tensor current | 3? |
| $\tilde{J}^i_\mu, \tilde{J}^i_{\mu 5}$ | $\mathrm{i}:\bar{q}\overleftrightarrow{\partial}_\mu(1, \mathrm{i}\gamma_5)q:$ | second-class hadronic vector and axial vector currents | 4? |
| $S^i_{\mu\nu}$ | $\mathrm{i}:\bar{q}(\gamma_\mu\overleftrightarrow{\partial}_\nu + \gamma_\nu\overleftrightarrow{\partial}_\mu)\frac{\lambda^i}{2}q:$ | symmetric tensor current | 4? |

Some comments should be made on the above list. First, the unit operator $I$ takes care of all $c$-number contributions to $\{O_n\}$. Second, the asymptotic scale dimension $d_\theta$ of the stress–energy tensor $\theta_{\mu\nu}$ is fixed at 4. In the dilation-invariant skeleton theory the commutator (see Chapter 3)

$$[D, P_\mu] = \mathrm{i}P_\mu \tag{4.51}$$

(where $P_\mu = \int \mathrm{d}^3x\, \theta_{0\mu}$) implies that the scale dimension of $\theta_{0\mu}$ is 4. Moreover, in this theory

$$[D, M_{\mu\nu}] = 0, \tag{4.52}$$

so that Lorentz rotations leave the scale dimension unaltered. The same, however, enable us to cover all components $\theta_{\mu\nu}$ starting from $\theta_{0\mu}$ only. A similar argument can be advanced for the hadronic vector and axial vector currents $J^i_\mu, J^i_{\mu 5}$. The commutators of the SU(3) × SU(3) charge algebra (Adler and Dashen 1968)

$$[Q^i(x^0), Q^j(x^0)] = \mathrm{i}f^{ijk}Q^k(x^0),$$
$$[Q^i_5(x^0), Q^j_5(x^0)] = \mathrm{i}f^{ijk}Q^k(x^0)$$

imply that $Q^i(x^0)$, $Q^i_5(x^0)$ are dimensionless, i.e. that $J^i_0(x)$ and $J^i_{05}(x)$ have the scale dimension of 3. Once again, eqn (4.52) implies that the asymptotic scale dimension of $J^i_\mu(x)$ and $J^i_{\mu 5}(x)$ are that same number. However, we cannot make such strong statements for the other operators shown in the list whose canonical scale dimensions only are given. What is definitely established about the symmetry properties of these operators is not sufficient to pin down their actual scale dimensions. For instance, the SU(3) × SU(3) transformation properties (Gell-Mann and Nee'man 1964) of the operators $u^i$ and $v^i$ merely imply

$$\begin{aligned}
[Q^i(x^0), u^j(x)] &= \mathrm{i}f^{ijk}u^k(x),\\
[Q^i(x^0), v^j(x)] &= \mathrm{i}f^{ijk}v^k(x),\\
[Q_5^i(x^0), u^j(x)] &= -\mathrm{i}d^{ijk}v^k(x),\\
[Q_5^i(x^0), v^j(x)] &= \mathrm{i}d^{ijk}u^k(x).
\end{aligned} \tag{4.53}$$

Eqns (4.53) are homogeneous in $u^i$ and $v^i$ and give no information on their scale dimensions. However, if we accept the U(12) algebra of Dashen and Gell-Mann (Dyson 1966) among the 144 generators $J^i_\mu$, $J^i_{\mu 5}$, $u^i$, $v^i$, $A^i_{\mu\nu}$, the equality between actual and canonical scale dimensions of these operators follows from the commutation relations of that theory. No such arguments which may be applied to the remaining operators $\tilde{J}^i_\mu$, $\tilde{J}^i_{\mu 5}$, $S^i_{\mu\nu}$ of the list is known to the author.

*OPE with two electromagnetic currents*

We now consider the short-distance expansion of the product of two electromagnetic currents as a concrete realization of the above ideas. We can take this to be

$$J^{\mathrm{EM}}_\mu(x)J^{\mathrm{EM}}_\nu(0) \underset{x\to 0}{\simeq} \sum_n C^n_{\mu\nu}(x)O_n(0).$$

We wish to consider all possible candidates for $O_n$ in the present case (Bonora and Vendramin 1971). But first we need take into account the effect of various symmetry constraints. The SU(3) decomposition of the hadronic electromagnetic current is given by

$$J^{\mathrm{EM}}_\mu = J^3_\mu + \frac{1}{\sqrt{3}}J^8_\mu. \tag{4.54}$$

Eqn (4.54) and the SU(3) × SU(3) algebra of currents (Gell-Mann and Nee'man 1964) imply that the equal-time commutator $[J^{\mathrm{EM}}_0(x), J^{\mathrm{EM}}_\mu(x_0, \mathbf{y})]$ is a Schwinger term, i.e. it contains at least one space derivative of a delta function. This information leads to the constraint

$$\int \mathrm{d}^3x\{C^n_{0\mu}(x_0 - \mathrm{i}\varepsilon, \mathbf{x}) - C^n_{0\mu}(x_0 + \mathrm{i}\varepsilon, \mathbf{x})\} = 0. \tag{4.55}$$

Moreover, the conservation of the electromagnetic current implies that

$$\partial^\mu C_{\mu\nu} = 0.$$

Finally, the requirements of hermiticity and of invariance under charge conjugation and parity transformation have to be satisfied.

If $O_n$ is a spin-0 object, it can be the unit operator $I$. The scalar and pseudoscalar densities $u^i$ and $v^i$ are absent from the leading terms, which are invariant under the group SU(3) × SU(3). This is because they belong to the representation $(3, 3^*) \oplus (3^*, 3)$ of SU(3) × SU(3), while the currents $J^i_\mu$ belong to the representation $(8,1) \oplus (1,8)$; products of the latter cannot generate the former. Among the possibilities with unit spin, the hadronic vector currents $J^i_\mu$ cannot contribute. This is because contributions such as

$$\mathrm{i}(-g_{\mu\nu}\Box + \partial_\mu\partial_\nu)\{x^\rho(-x^2+\mathrm{i}\varepsilon x^0)^{-1}\}J^{3,8}_\rho,$$

$$\mathrm{i}\{(g_{\mu\alpha}g_{\nu\beta}+g_{\mu\beta}g_{\nu\alpha})\Box - (g_{\mu\alpha}\partial_\nu + g_{\nu\alpha}\partial_\mu)\partial_\beta - \\ -(g_{\mu\beta}\partial_\nu + g_{\nu\beta}\partial_\mu)\partial_\alpha + 2g_{\mu\nu}\partial_\alpha\partial_\beta\}\{x^\rho x^\alpha x^\beta(-x^2+\mathrm{i}\varepsilon x^0)^{-2}\}J^{3,8}_\rho,$$

which are the only forms compatible with the other symmetry constraints, make non-zero non-Schwinger contributions to the commutator in the equal-time limit, and the corresponding $c$-number coefficients $C^\rho_{\mu\nu}$ violate eqn (4.55). The axial vector currents $J^i_{\mu 5}$ can contribute a term of the type

$$C^{3,8}_{\mu\nu\alpha}(x)J^{\alpha,3,8}_5 = C^{3,8}\varepsilon_{\mu\nu\alpha\lambda}\partial^\lambda(-x^2+\mathrm{i}\varepsilon x^0)^{-1}J^{\alpha,3,8}_5,$$

consistent with all the above constraints. Any contribution from second-class currents can be ruled out, since they belong to the representation $(3,3^*) \oplus (3^*,3)$ of SU(3) × SU(3). Now we come to operators with spin-2. Here the stress–energy tensor $\theta_{\mu\nu}$ can certainly contribute. The antisymmetric tensor currents $A^i_{\mu\nu}$ cannot contribute, owing to wrong C-conjugation properties. However, the symmetric tensor currents $S^i_{\mu\nu}$ can contribute. For the present, let us not consider operators with higher spins. We thus obtain the expansion

$$J^{\mathrm{EM}}_\mu(x)J^{\mathrm{EM}}_\nu(0) \simeq C_{\mu\nu}(x)I + C^3_{\mu\nu\alpha}(x)J^\alpha_{5,3}(0) + C^8_{\mu\nu\alpha}(x)J^{\alpha,8}_5(0) + C^0_{\mu\nu\alpha\beta}(x)S^{\alpha\beta,0}(0) + \\ + C^3_{\mu\nu\alpha\beta}(x)S^{\alpha\beta,3}(0) + C^8_{\mu\nu\alpha\beta}(x)S^{\alpha\beta,8}(0) + \ldots \tag{4.56}$$

In eqn (4.56), we have used the notation $S^{\alpha\beta,0}$ for the stress–energy tensor $\theta^{\alpha\beta}$ and tentatively taken $S^{\alpha\beta,3}$, $S^{\alpha\beta,8}$ to have their canonical scale dimension of 4. In addition, the $c$-number coefficients are given by the following relations:

$$C_{\mu\nu}(x) = A(-g_{\mu\nu}\Box + \partial_\mu\partial_\nu)(-x^2+\mathrm{i}\varepsilon x_0)^{-2}, \tag{4.57}$$

$$C^{3,8}_{\mu\nu}(x) = C^{3,8}\varepsilon_{\mu\nu\alpha\lambda}\partial^\lambda(-x^2+\mathrm{i}\varepsilon x_0)^{-1}, \tag{4.58}$$

$$C^{0,3,8}_{\mu\nu\alpha\beta}(x) = B^{0,3,8}_{\mathrm{L}}(-g_{\mu\nu}\Box + \partial_\mu\partial_\nu)\{x_\alpha x_\beta(-x^2+\mathrm{i}\varepsilon x_0)^{-1}\} + \\ + B^{0,3,8}_2(-g_{\mu\alpha}g_{\nu\beta}\Box + g_{\nu\beta}\partial_\mu\partial_\alpha + g_{\mu\alpha}\partial_\nu\partial_\beta - 2g_{\mu\nu}\partial_\alpha\partial_\beta - g_{\nu\alpha}g_{\mu\beta}\Box + \\ + g_{\mu\beta}\partial_\nu\partial_\alpha + g_{\nu\alpha}\partial_\mu\partial_\beta)\tfrac{1}{2}\ln(-x^2+\mathrm{i}\varepsilon x^0). \tag{4.59}$$

In eqn (4.56) we have displayed only those terms in the OPE that are of immediate interest to us. Current conservation and scale invariance are seen to be manifest in these terms when the information contained in eqns (4.57)–(4.59) is used.

**Applications of the Wilson framework**

In discussing the applications of the Wilson framework we shall confine ourselves to deep inelastic electromagnetic processes only. However, those of our remarks which pertain to scattering may be generalized in a straightforward way to include the corresponding reactions induced by weak interactions. Our starting point is eqn (4.56). On the right-hand side of that equation the disconnected $C_{\mu\nu}(x)I$ term, which scales as $x^{-6}$, finds application in $e^+e^-$ annihilation. The terms with $C_{\mu\nu\alpha}(x)$, going as $x^{-3}$, are applicable to spin-dependent $l$N scattering in the deep inelastic domain. Those with $C_{\mu\nu\alpha\beta}(x)$ proportional to $x^{-2}$ are important for spin-averaged deep inelastic scattering. We consider only the first and the third cases here. The second case may be treated as the third one, and in any event we shall return to the topic of spin-dependence in deep inelastic eN scattering in Part III.

*High-energy $e^+e^-$ annihilation*

We shall consider the reaction $e^+e^- \to$ 'anything' to which we first alluded in the Introduction. To the lowest order in electromagnetism, the single-photon exchange diagram may be drawn as in Fig. 35. First rewrite eqn (4.57) as

$$C_{\mu\nu}(x) = -12A(g_{\mu\nu}x^2 - x_\mu x_\nu)(-x^2 + i\varepsilon x_0)^{-4}. \tag{4.60}$$

We now take the $S$-matrix

$$S = \left(\frac{1}{2\pi}\right)^3 \frac{m_e}{\sqrt{(k_{1,0}k_{2,0})}} i(2\pi)^4 \frac{\delta^{(4)}(k_1 + k_2 - q)}{q^2 + i\varepsilon} \bar{v}(k_2)\gamma^\mu u(k_1) \langle n|J_\mu|0\rangle, \tag{4.61}$$

square the same, and use the relation

$$\sum_n \langle 0|J_\mu^{\mathrm{EM}}|n\rangle \langle n|J_\nu^{\mathrm{EM}}|0\rangle \delta^{(4)}(q - p_n)(2\pi)^4$$
$$= \int d^4x\, e^{iq.x} \langle 0|J_\mu^{\mathrm{EM}}(x) J_\nu^{\mathrm{EM}}(0)|0\rangle \tag{4.62}$$

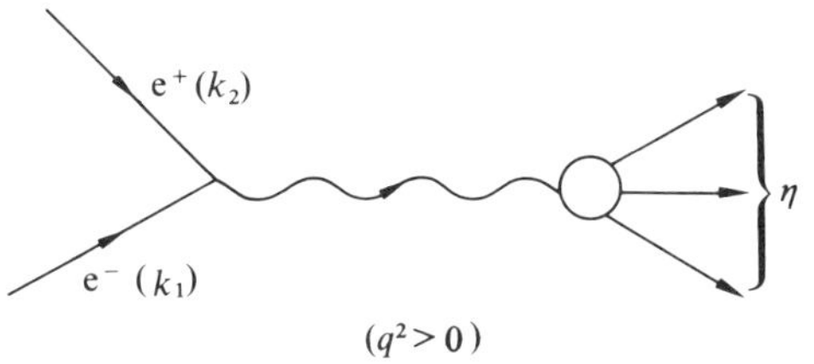


FIG. 35. Lowest-order (electromagnetic) $e^+e^-$ annihilation into hadrons.

to obtain the total cross-section as a function of $q^2$:

$$\sigma_{\text{tot}}^{e^+e^-}(q^2) = -\frac{16}{3}\frac{\pi^2\chi^2}{q^4}\int d^4x\, e^{iq.x}\langle 0|J_\mu^{\text{EM}}(x)J^{\text{EM},\mu}(0)|0\rangle. \tag{4.63}$$

Let us investigate $\sigma_{\text{tot}}(q^2)$ as $q^2 \to \infty$. Go to the $\mathbf{q} = 0$ frame, i.e. to the centre-of-mass frame of the initial leptons. Since $q_0 > 0$, we can replace the ordinary product $J_\mu^{\text{EM}}(x)J^{\text{EM},\mu}(0)$ in eqn (4.63) by the commutator $[J_\mu^{\text{EM}}(x), J^{\text{EM},\mu}(0)]$. Thus we need to consider the quantity

$$\lim_{q^0\to\infty}\int_{-\infty}^{\infty} dx^0 \exp(iq^0x^0)\int d^3x\,[J_\mu^{\text{EM}}(x), J^{\text{EM},\mu}(0)]. \tag{4.64}$$

By the Riemann–Lebesgue theorem, only times corresponding to $x^0 \to 0$ contribute to the integrand of the expression in eqn (4.64) as $q_0 \to \infty$. Because of the micro-causality property of the commutator, this implies that $\mathbf{x} \to 0$. In other words only the region near $x^\mu = 0$ contributes to the said integrand. That is just the tip of the light cone. Since we take the vacuum expectation value of the commutator, only the disconnected $c$-number term in the Wilson expansion contributes. This leads us from eqn (4.63) to

$$\lim_{q^2\to\infty}\sigma_{\text{tot}}^{e^+e^-}(q^2) = -\frac{16}{3}\frac{\pi^2\alpha^2}{q^4}\lim\int d^4x \exp(iq^0x^0)\, C_\mu^\mu(x). \tag{4.65}$$

It follows from eqn (4.65) that (see Exercise 4.3)

$$\lim_{q^2\to\infty}\sigma_{\text{tot}}^{e^+e^-}(q^2) = -\frac{32\pi^5A\alpha^2}{q^2}. \tag{4.66}$$

If scale invariance is assumed, the $q^{-2}$ dependence of $\sigma_{\text{tot}}^{e^+e^-}$ in the high-$q^2$ limit is rather trivially obtained. The only four-momentum involved being $q$, an ordinary dimensional analysis yields that $\sigma_{\text{tot}}^{e^+e^-} = cq^{-2}$, $c$ being a dimensionless constant. If experiment shows that $\sigma_{\text{tot}}^{e^+e^-}$ approaches zero more strongly than $1/q^2$ as $q^2 \to \infty$, in eqn (4.66) $A$ is zero for some unknown dynamical reason; we then have to consider scale-breaking contributions. If $\sigma_{\text{tot}}^{e^+e^-}$ decreases more slowly than $1/q^2$ in that limit, a breakdown in the idea of scale invariance at short distances is unambiguously implied. It is emphasized, however, that the onset of the scale invariant behaviour in the time-like region covered by this process need not be as precocious as observed in inelastic eN scattering (see also Part III, p. 148).

*Spin-averaged deep inelastic eN scattering*

We shall consider this reaction by combining the idea of short-distance scale invariance with the technique of dispersion relations (Wilson 1971a; Mack 1971; Ciccariello, Gatto, Sartori, and Tonin 1971). Because of spin-averaging, the axial vector currents $J^{3;8}_{\mu 5}$ do not contribute from the right-hand side of eqn (4.56). This is because the matrix element $\langle N|J^{3;8}_{\mu 5}|N\rangle$ vanishes when nucleon spins are averaged. Hence in the OPE of eqn (4.56), the next connected terms with the coefficients $C^{0,3,8}_{\mu\nu\alpha\beta}$ need be considered. First, we return to eqn (2.29) and write

$$T^{\mathrm{e}}_{\mu\nu} = (2\pi)^2 \frac{2\mathrm{i}p_0}{M} \int \mathrm{d}^4x\, \mathrm{e}^{\mathrm{i}q.x}\theta(x^0)\langle \mathrm{N}|[J^{\mathrm{EM}}_{\mu}(x), J^{\mathrm{EM}}_{\nu}(0)]|\mathrm{N}\rangle + R_{\mu\nu}(q),$$
$$W^{\mathrm{e}}_{\mu\nu} = \mathrm{Im}\, T^{\mathrm{e}}_{\mu\nu}. \tag{4.67}$$

In eqn (4.67) $R_{\mu\nu}$ is the seagull term (with real polynomial coefficients). The decomposition of $T^{\mathrm{e}}_{\mu\nu}$, consistent with current conservation, is as follows:

$$T^{\mathrm{e}}_{\mu\nu} = \frac{\nu}{q^2}\left(-g_{\mu\nu} + \frac{q_\mu q_\nu}{q^2}\right) T^{\mathrm{e}}_{\mathrm{L}}(q^2, \nu) + \frac{1}{M^2}\left\{q^2 p_\mu p_\nu - q\,.\,p(q_\mu p_\nu + q_\nu p_\mu)\right.$$
$$\left. + g_{\mu\nu}(q\,.\,p)^2\right\}\frac{1}{q^2} T^{\mathrm{e}}_{2}(q^2, \nu). \tag{4.68}$$

In eqn (4.68)

$$T^{\mathrm{e}}_{\mathrm{L}} = \frac{q^2}{\nu} T^{\mathrm{e}}_{1} + \nu T^{\mathrm{e}}_{2}. \tag{4.69}$$

Further, we have $\mathrm{Im}\, T^{\mathrm{e}}_{1,2,\mathrm{L}} = W^{\mathrm{e}}_{1,2,\mathrm{L}}$, $W^{\mathrm{e}}_{\mathrm{L}}$ being $\nu W^{\mathrm{e}}_{2} - 2MwW^{\mathrm{e}}_{1}$.

The analyticity properties of $T^{\mathrm{e}}_{1}$ and $T^{\mathrm{e}}_{2}$ in $\nu$ are now needed. This is because the OPE gives only the $x^\mu \to 0$ behaviour, whereas the reaction in the deep inelastic region turns out to probe all of the light cone (see Part III, p. 106). We assume that the total amplitude for forward virtual Compton scattering has pomeron-exchange behaviour in the Regge limit. This means that, when $\nu \to \infty$ with $q^2$ fixed, the amplitude develops the form $\beta(q^2)\nu$. Summing over polarizations, we have

$$\sum_{\varepsilon} \varepsilon^\mu \varepsilon^\nu T^{\mathrm{e}}_{\mu\nu} = \left(-g_{\mu\nu} + \frac{q_\mu q_\nu}{q^2}\right) T^{\mathrm{e}}_{\mu\nu}$$
$$= 3T^{\mathrm{e}}_{1} - \left(1 - \frac{\nu^2}{q^2}\right) T^{\mathrm{e}}_{2}$$
$$\underset{\mathrm{Regge}}{\simeq} \beta(q^2)\nu,$$

($\beta(q^2)$ being an arbitrary function of $q^2$). Hence, it follows (with $t_{1,2}(q^2)$ being arbitrary functions of $q^2$) that

$$\lim_{\mathrm{Regge}} T^{\mathrm{e}}_{1}(q^2, \nu) = t_1(q^2)\nu, \tag{4.70a}$$

$$\lim_{\mathrm{Regge}} T^{\mathrm{e}}_{2}(q^2, \nu) = t_2(q^2)\nu^{-1}. \tag{4.70b}$$

Thus we may take a subtracted dispersion relation for $T_1$ and an unsubtracted dispersion relation for $T_2$. For $q^2 < 0$, the presence of sufficient analyticity in the $\nu$-plane for the existence of these dispersion relations can be established. Thus we can write

$$T_2^e(q^2, \nu) = \frac{1}{\pi} \int_{-\infty}^{\infty} d\nu' \frac{W_2^e(q^2, \nu')}{\nu' - \nu + i\varepsilon}$$

$$= \frac{1}{\pi} \int_{\nu_{\min}}^{\infty} d\nu' \left( \frac{1}{\nu' - \nu - i\varepsilon} + \frac{1}{\nu' + \nu + i\varepsilon} \right) W_2^e(q^2, \nu),$$

using the crossing property of $W_2^e(q^2, \nu)$—as obtained from eqn (I.12)—in the last step. We have then

$$T_2^e(q^2, \nu) = \frac{1}{\pi} \int_{\nu_{\min}}^{\infty} d\nu'^2 \frac{W_2^e(q^2, \nu')}{\nu'^2 - \nu^2 - i\varepsilon}. \tag{4.71}$$

The amplitude $T_1$ needs two subtractions and can be written as

$$T_1^e(q^2, \nu) = C(q^2)\nu + T_1^e(q^2, 0) + \frac{\nu^2}{\pi} \int_{-\infty}^{\infty} \frac{d\nu'}{\nu'^2} \frac{W_2^e(q^2, \nu')}{\nu' - \nu - i\varepsilon}.$$

The evenness of $T_1^e$ with respect to $q$, i.e. the relation $T_1^e(q^2, \nu) = T_1^e(q^2, -\nu)$, implies that $C$ has to vanish identically. Hence we may write

$$T_1^e(q^2, \nu) = T_1^e(q^2, 0) + \frac{\nu^2}{\pi} \int_0^{\infty} \frac{d\nu'^2}{\nu'^2} \frac{W_1^e(q^2, \nu')}{\nu'^2 - \nu^2 - i\varepsilon}. \tag{4.72}$$

In terms of the scale variable $w = -q^2/2M\nu$, the dispersion relations (eqns (4.71) and (4.72)) read

$$T_2^e(q^2, w) = -\frac{w^2}{\pi} \int_0^1 \frac{dw'^2}{w'^2} \frac{W_2^e(q^2, w')}{w'^2 - w^2 - i\varepsilon},$$

$$T_1^e(q^2, w) = T_1^e(q^2, \infty) - \frac{1}{\pi} \int_0^1 dw'^2 \frac{W_1^e(q^2, w')}{w'^2 - w^2 - i\varepsilon}. \tag{4.73}$$

At this point, we take the limit $q^2 \to -\infty$ in the following special way. Put $q = \lambda^{-1}\hat{q}$, where $\hat{q}$ is the space-like vector $(i, \mathbf{0})$ and take the 'dilational limit' (Mack 1971), where $\lambda$ approaches zero. Thus $q^2 = -1/\lambda^2$ tends to $-\infty$. Eqn (4.67) may now be rewritten as

$$T^{\text{e}}_{\mu\nu} = (2\pi)^2 \frac{2\text{i}p_0}{M} \int_0^\infty \text{d}x^0 \exp(-x^0/\lambda) \int \text{d}^3x \langle \text{N}|[J^{\text{EM}}_\mu(x), J^{\text{EM}}_\nu(0)]|\text{N}\rangle$$

$$+ R_{\mu\nu}(q). \tag{4.74}$$

Moreover, in this limit, $w \to 1/(2p_0\text{i}\lambda)$, so that

$$T^{\text{e}}_2 \to \frac{2}{\pi} \int_0^1 \frac{\text{d}w'}{w'} W^{\text{e}}_2(-\lambda^{-2}, w'), \tag{4.75a}$$

$$T^{\text{e}}_1 \to T^{\text{e}}_1(-\lambda^{-2}, \infty) - \frac{8}{\pi} p_0^2 \lambda^2 \int_0^1 \text{d}w'\, w' W^{\text{e}}_1(-\lambda^{-2}, w'). \tag{4.75b}$$

Moreover, in the same limit it follows from eqn (4.69) that

$$T^{\text{e}}_{\text{L}} \to -\frac{T^{\text{e}}_1 M}{\text{i}p_0\lambda} + \frac{\text{i}p_0}{M\lambda} T^{\text{e}}_2. \tag{4.76}$$

As $\lambda \to 0$, according to the Riemann–Lebesgue theorem, only the $x^0 \to 0$ region is picked out in eqn (4.74). This forces $\mathbf{x} \to 0$ in the casual commutator of that equation. Thus the 'dilational limit' does probe the commutator of eqn (4.74) exclusively at short distances, enabling us to use the OPE of eqn (4.56). Let us continue for the present to have the fiction that only operators of spin up to 2 need be considered in that OPE. Choosing $\mu, \nu = i, j$ (space-like), we then have

$$\lim_{\lambda \to 0} T^{\text{e}}_{ij} = 2(2\pi)^2 \frac{\text{i}p_0}{M} \sum_{k=0,3,8} \int_0^\infty \text{d}x_0 \exp(-x_0/\lambda) \int \text{d}^3x \times$$

$$\times \{C^k_{ij\alpha\beta}(x_0 - \text{i}\varepsilon, \mathbf{x}) - C^k_{ij\alpha\beta}(x_0 + \text{i}\varepsilon, \mathbf{x})\} \langle \text{N}|S^{\alpha\beta,k}(0)|\text{N}\rangle +$$

$$+ R(\lambda)\delta_{ij}. \tag{4.77}$$

In general, we can write the forward spin-averaged nucleon matrix element of the tensor current as

$$(2\pi)^3 \frac{p_0}{M} \langle \text{N}|S^{\alpha\beta,k}(0)|\text{N}\rangle = \frac{p^\alpha p^\beta}{M} S^k_{\text{N}} + g^{\alpha\beta} C^k_{\text{N}}, \tag{4.78}$$

where $S^k_{\text{N}}$, $C^k_{\text{N}}$ are constants with $S^0_{\text{N}} = 1$, $C^0_{\text{N}} = 0$ by virtue of the universality of the stress–energy tensor. Thus, using eqns (4.59) and (4.78) in eqn (4.77) and retaining only the leading terms, we have

$$\lim_{\lambda\to 0} T^{\rm e}_{ij} = \frac{1}{\pi}\sum_{k=0,3,8}\int_0^\infty {\rm d}x_0 \exp(-x_0/\lambda)\int {\rm d}^3x\,[B^k_{\rm L}S^k_{\rm N}(\delta_{ij}\Box+\partial_i\partial_j)x_\alpha x_\beta \times$$
$$\times\{(-x^2+{\rm i}\varepsilon x_0)^{-1}-(-x^2-{\rm i}\varepsilon x_0)^{-1}\}\frac{p^\alpha p^\beta}{M}+$$
$$+B^k_2S^k_{\rm N}\{-p_ip_j\Box-p\,.\,\partial(p_j\partial_i+p_i\partial_j)+\delta_{ij}(p\,.\,\partial)^2\}\times$$
$$\times\{\ln(-x^2+{\rm i}\varepsilon x_0)-\ln(-x^2-{\rm i}\varepsilon x_0)\}]+R(\lambda)\delta_{ij}. \tag{4.79}$$

Note that the $g_{\alpha\beta}$ terms give lesser singularities and hence have been ignored. From purely dimensional arguments we may write

$$\int {\rm d}^4x\,{\rm e}^{{\rm i}q.x}(-g_{\mu\nu}\Box+\partial_\mu\partial_\nu)\frac{x_\alpha x_\beta}{x^2}$$
$$\propto\frac{1}{q^2}\left(g_{\mu\nu}-\frac{q_\mu q_\nu}{q^2}\right)\left(-g_{\alpha\beta}+\frac{4q_\alpha q_\beta}{q^2}\right),$$
$$\int {\rm d}^4x\,{\rm e}^{{\rm i}q.x}\{-p_\mu p_\nu\Box+p\,.\,\partial(p_\mu\partial_\nu+p_\nu\partial_\mu)-g_{\mu\nu}(p\,.\,\partial)^2\}\ln(-x^2)$$
$$\propto\frac{1}{q^2}\{-q^2p_\mu p_\nu+(p\,.\,q)(p_\mu q_\nu+p_\nu q_\mu)-g_{\mu\nu}(p\,.\,q)^2\}.$$

Using these facts in eqn (4.79), we obtain

$$\lim_{\lambda\to 0} T^{\rm e}_{ij} = \sum_{k=0,3,8}\frac{B^{k\prime}_{\rm L}S^k_{\rm N}\lambda^2}{M}\delta_{ij}(4p_0^2-M^2)+$$
$$+4\sum_{k=0,3,8}\frac{B^{k\prime}_2S^k_{\rm N}\lambda^2}{M}(\delta_{ij}p_0^2-p_ip_j)+R(\lambda)\delta_{ij}, \tag{4.80}$$

where the $B^{k\prime}_{\rm L,2}$ are constants proportional to $B^k_{\rm L,2}$. On the other hand, eqn (4.68) has now become

$$\lim T^{\rm e}_{ij} = \delta_{ij}\lim\left(T^{\rm e}_1+\frac{p_0^2}{M^2}T^{\rm e}_2\right)-(\delta_{ij}p_0^2-p_ip_j)\frac{1}{M^2}\lim T^{\rm e}_2$$
$$=\delta_{ij}\left\{T^{\rm e}_1(-\lambda^2,\infty)-\frac{8}{\pi}p_0^2\lambda^2\int_0^1{\rm d}w'\,w'W^{\rm e}_1(-\lambda^2,w')+\right.$$
$$\left.+\frac{2p_0^2}{\pi M^2}\int_0^1\frac{{\rm d}w'}{w'}W^{\rm e}_2(-\lambda^{-2},w')\right\}+$$
$$+\frac{2}{\pi M^2}(\delta_{ij}p_0^2-p_ip_j)\int_0^1\frac{{\rm d}w'}{w'}W^{\rm e}_2(-\lambda^{-2},w'), \tag{4.81}$$

the last step following by use of eqn (4.75). Using the definition of $W^e_L$, eqn (4.81) may be rewritten as

$$\lim T^e_{ij} = \delta_{ij}\left\{T^e_1(-\lambda^{-2}, \infty)+\frac{4\lambda^2}{M\pi}p_0^2\int_0^1 dw\, W^e_L(-\lambda^{-2}, w)\right\}+$$

$$+\frac{4\lambda^2}{M\pi}(\delta_{ij}p_0^2-p_ip_j)\int_0^1 dw\, \nu W^e_2(-\lambda^{-2}, w). \tag{4.82}$$

Comparing the coefficients of the tensors in the two expressions of eqns (4.80) and (4.82), we obtain

$$T^e_1(-\lambda^{-2}, \infty) = -\sum_{k=0,3,8}\frac{B^{k\prime}_L S^k_N M}{\lambda^2}+R(\lambda), \tag{4.83a}$$

$$\lim_{\lambda\to 0}\int_0^1 dw\, W^e_L(-\lambda^{-2}, w) = \pi\sum_{k=0,3,8} B^{k\prime}_L S^k_N, \tag{4.83b}$$

$$\lim_{\lambda\to 0}\int_0^1 dw\, \nu W^e_2(-\lambda^{-2}, w) = \pi\sum_{k=0,3,8} B^{k\prime}_2 S^k_N. \tag{4.83c}$$

The last two integrals are called the Callan–Gross integrals. They imply the scaling property that, for $q^2 \to -\infty$ and $w$ fixed, $MW^e_1 \to F^e_1(w)$ and $\nu W^e_2 \to F^e_2(w)$, where the $F^e_{1,2}(w)$ are some functions of $w$. Thus from the Wilson expansion plus dispersion relations in the $\nu$-plane we have come to Bjorken scaling for the inelastic eN structure functions. The crucial fact behind this is that the contribution to the OPE came from spin-2 operators of asymptotic scale dimension 4. $S^0_{\alpha\beta}$, i.e. $\theta_{\alpha\beta}$, of course, must have that dimension. However, $S^{3,8}_{\alpha\beta}$ (or any other spin-2 operator which may contribute) must not have anything less. Indeed, it may be shown (see Exercise 4.4) that the contributions to $F^e_{1,2}$ from spin-2 fields in the OPE with asymptotic scale dimensions greater or less than 4 tend to 0 or $\infty$ respectively. Now, if the stress–energy tensor $\theta_{\alpha\beta}$ were the only field of spin-2 contributing to the present matrix element of the OPE with $d \leqslant 4$, then the summation over $k$ in eqn (4.83) would have covered only $k = 0$. Because $S^0_N = 1$, this would have led to

$$\pi B^{0\prime}_L = \int_0^1 dw\, F^e_2(w) = \int_1^\infty \frac{d\omega}{\omega^2}F^e_2(\omega); \tag{4.84}$$

in other words the integral of eqn (4.84) would have been universal (i.e. target-independent). This is the Mack sum rule. Experimentally (Kendall 1971), for proton and neutron (the latter in deuterium) targets, the situation

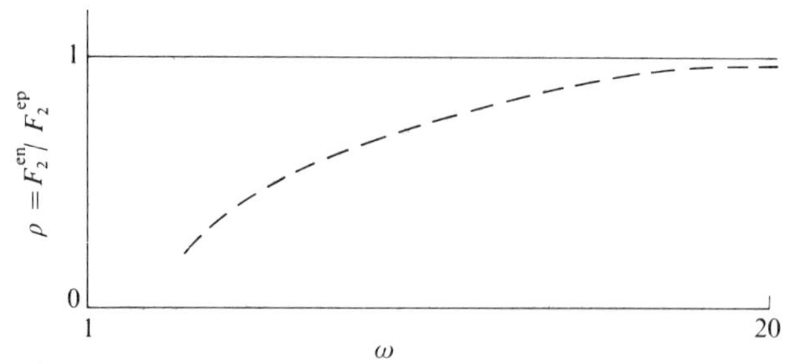


FIG. 36. Simplified representation of the experimental behaviour of the ratio of the structure functions $F_2$ for electron–neutron and electron–proton interactions (dashed line).

is as illustrated in Fig. 36 (cf. Fig. 20). Clearly, the Mack sum rule is wrong. The experimental curve, if it is not strictly unity, should oscillate around that value in order to satisfy the sum rule. The observation to the contrary proves the presence of spin-2 fields with $d = 4$, in particular that of the tensor fields $S^{3,8}_{\mu\nu}$ in the OPE of eqn (4.56), with their asymptotic scale dimensions equal to the canonical value 4. Thus the failure of the Mack sum rule may be regarded as a setback for the anomalous dimension hypothesis.

The above discussion has been made particularly simple by our summary disregard of fields with spins higher than 2 contributing to eqn (4.56). This procedure was not quite correct, since higher-spin fields probably contribute to the OPE. Instead of eqn (4.56), we may consider the more general form of the OPE as

$$J^{\mathrm{EM}}_{\mu}(x)J^{\mathrm{EM}}_{\nu}(0) \underset{x\to 0}{\simeq} \sum_{m}\sum_{J} C^{(m)\alpha_1\ldots\alpha_J}_{\mu\nu}(x)O^{(m)}_{\alpha_1\ldots\alpha_J}(0) \tag{4.85}$$

and trace the earlier argument with it. In eqn (4.85) $O_{\alpha_1\ldots\alpha_J}$ is a field of spin $J$ and asymptotic scale dimension $d_J$; the superscript $(m)$ takes into account all internal indices and $C^{(m)}(x) \propto x^{-6+d_J}$. In the dilational limit, eqn (4.77) will now be changed to

$$\begin{aligned}\lim_{\lambda\to 0} T^{\mathrm{e}}_{\mu\nu} = 2(2\pi)^2\frac{\mathrm{i}p_0}{M}\sum_{J}\sum_{m}\int_0^{\infty}\mathrm{d}x^0\exp(-x^0/\lambda)\int\mathrm{d}^3x\\ \times\left\{C^{(m)\alpha_1\ldots\alpha_J}_{\mu\nu}(x_0-\mathrm{i}\varepsilon,\mathbf{x})-C^{(m)\alpha_1\ldots\alpha_J}_{\mu\nu}(x_0+\mathrm{i}\varepsilon,\mathbf{x})\right\}\\ \times\langle\mathrm{N}|O_{\alpha_1\ldots\alpha_J}(0)|\mathrm{N}\rangle + R(\lambda)g_{\mu\nu}.\end{aligned} \tag{4.86}$$

Only fields which are completely symmetric in the Lorentz indices will contribute to the spin-averaged forward nucleon matrix element of eqn (4.86). This contribution may be written as

$$(2\pi)^3\frac{p_0}{M}\langle\mathrm{N}|O^{(m)}_{\alpha_1\ldots\alpha_J}(0)|\mathrm{N}\rangle = S^{(m)}_{\mathrm{N}}p_{\alpha_1}\cdots p_{\alpha_J} + \text{'trace term'}. \tag{4.87}$$

The 'trace term' here indicates something involving one or more of the metric tensors $g_{\alpha_i\alpha_k}$. Again this makes a non-leading contribution in the limit of interest, and may be ignored. Thus the constants $\{S_N^{(m)}\}$ play the same sort of role as the $S_N^{0,3,8}$ introduced earlier. We can now follow through with the rest of the argument. The procedure is very similar to that above, but tedious technicalities are involved. We will not discuss the details here, but they may be found in the literature (Mack 1971). The final result is that, in place of eqn (4.83), we obtain relations with moments of Calla–Gross integrals, namely,

$$\lim_{\lambda\to 0}\int_0^1 \mathrm{d}w\, w^{J-2} W^{\mathrm{e}}_{\mathrm{L},J}(-\lambda^{-2}, w) = \pi \sum_m B_{\mathrm{L}}^{(m)\prime} S_{\mathrm{N}}^{(m)}(\lambda)^{d_J - J - 2}, \qquad (4.88\mathrm{a})$$

$$\lim_{\lambda\to 0}\int_0^1 \mathrm{d}w\, w^{J-2} \nu W^{\mathrm{e}}_{2,J}(-\lambda^{-2}, w) = \pi \sum_m B_{2}^{(m)\prime} S_{\mathrm{N}}^{(m)}(\lambda)^{d_J - J - 2}, \qquad (4.88\mathrm{b})$$

where $W^{\mathrm{e}}_{i,J}$ denotes the contribution to $W^{\mathrm{e}}_i$ from a field $O_{\alpha_1\ldots\alpha_J}$ of spin $J$ and asymptotic scale dimension $d_J$ in the OPE and $B_{\mathrm{L},2}^{(m)\prime}$ are constants analogous to the $B_{\mathrm{L}}^{0,3,8\prime}$ of eqn (4.83). However, Bjorken scaling for $W^{\mathrm{e}}_{\mathrm{L}}$ and $\nu W^{\mathrm{e}}_2$ would require all these moment integrals to either vanish or be finite. Hence the asymptotic scale dimensions of the fields $O_{\alpha_1\ldots\alpha_J}$ must obey the constraint

$$d_J \geqslant J+2 \qquad (\text{for } J > 2). \qquad (4.89)$$

This inequality is necessary for Bjorken scaling, which thus cannot be derived from scale invariance at short distances and dispersion relations alone. The canonical value for $d_J$ is $J+2$, so that Bjorken scaling does emerge naturally in the hypothesis of Brandt and Preparata. Otherwise, eqn (4.89) may be regarded as a restriction on the anomalous dimensions of fields with $J > 2$.

Of course, the canonical nature of the stress tensor ensures that there is at least one term in each of the sums of eqns (4.88) which is non-zero when $\lambda \to 0$; thus the moment integrals are all non-vanishing in this limit.

## Exercises

4.1. Justify the structure of the short-distance singularity for the time-ordered product of two operators, as claimed in eqn (4.48).

4.2. Establish the connection in between short-distance OPE and Schwinger terms in the equal-time commutator as given in eqns (4.49) and (4.50).

4.3. Derive eqn (4.66) for the $e^+e^-$ total hadronic annihilation cross-section.

4.4. Show that the contributions from spin-2 fields with asymptotic scale dimensions greater or less than 4 in the OPE of eqn (4.56) to Bjorken scaling vanish or diverge respectively.

# PART III
# LIGHT-CONE PHYSICS

# 5
# RUDIMENTS OF LIGHT-CONE PHYSICS

## Preliminaries

THE fundamental importance of the light-cone $x^2 = 0$ is derived from the micro-causality principle. This law decrees that two spatially separated local measurements cannot affect each other during any time interval that is less than that needed by light to cover the distance of separation. Consequently, the commutator of the corresponding local operators vanishes outside the light-cone. Since, in general, it is nonvanishing inside, the existence of light-cone singularities in commutators and, more broadly, in operator products is immediately implied. For certain products, such singularities are measurable in terms of inelastic lepton–hadron processes involving large momentum transfers. This is only to be expected. With $q$ and $x$ being conjugate variables, it is but natural that the limit $q^2 \to \infty$ should yield information around $x^2 = 0$. The available data on deep inelastic lepton–hadron scattering suggest that the leading structure of the light-cone singularities in commutators of hadronic currents is that of free quark-field theory (Frishman 1971a,b; Brandt and Preparata 1971). Suppose we make this the starting point of our discussions of light-cone physics. Of course, we can be a little more sophisticated; for instance, we may use the formal field theory of quarks and gluons rather than that of free quarks. However, this does not alter the substance of the statement made earlier (Llewellyn-Smith 1971). With that postulate we are led to various explanations of observed phenomena, as well as to new predictions. Further, we are rewarded with a Lie algebra of currents on the light-cone of considerable mathematical elegance (Fritzsch and Gell-Mann 1971a,b).

First, we shall establish more precisely the relevance of the light-cone to deep inelastic $l$N scattering. Then we shall give an example of a light-cone singularity from free field theory before taking up the topic of light-cone expansions. The latter will be followed by discussions of light-cone current algebra and of the light-plane approach.

### *Light-cone dominance*

The validity of light-cone dominance in the commutator of two hadronic weak or electromagnetic currents is a consequence of the Riemann–Lebesgue lemma (Lighthill 1964) plus some very plausible assumptions. Consider the structure tensors of eqns (I.10), (I.23), and (I.25).

$$W_{\mu\nu} = (2\pi)^2 \frac{p_0}{M} \int \mathrm{d}^4x\, \mathrm{e}^{iq.x} \langle \mathrm{N}|[J_\mu^\dagger(x), J_\nu(0)]|\mathrm{N}\rangle, \tag{5.1}$$

where $J_\mu$ is an appropriate hadronic current. In the laboratory frame, the energy transfer $q_0 = \nu \equiv M^{-1}p.q$. Choose the unit vector $\hat{q} = \hat{x}_3$. Then $q_3 = \sqrt{(\nu^2+Q^2)}$ and

$$\lim_{\mathrm{Bj}} q_3 = \nu + \frac{Q^2}{2\nu} + \cdots = \nu + Mw + O\left(\frac{M^2w^2}{\nu}\right). \tag{5.2}$$

Change the integration variables from $(x^0, x^3)$ to $(x^0 - x^3, x^3)$. Then eqn (5.1) may be rewritten as

$$\lim_{\mathrm{Bj}} W_{\mu\nu} = \int \mathrm{d}(x^0 - x^3) \int \mathrm{d}x^3 \int \mathrm{d}^2x_\perp \exp\{\mathrm{i}\nu(x^0 - x^3)\} \times$$
$$\times \exp(\mathrm{i}Mwx^3) f_{\mu\nu}(x, p), \tag{5.3}$$

where

$$f_{\mu\nu}(x, p) = \langle \mathrm{N}|[J_\mu^\dagger(x), J_\nu(0)]|\mathrm{N}\rangle. \tag{5.4}$$

We first give a naive argument. As $\nu \to \infty$, the contributions to the integrals come mainly from the region $x^0 - x^3 \simeq O(1/\nu)$, $x^3$ finite. (We assume, of course, that the matrix element has no infinitely large dynamical phase such as $\exp\{-\mathrm{i}m(x^0 - x^3)\}$—with $m$ being a very large mass—which could compensate the factor $\exp\{\mathrm{i}\nu(x^0 - x^3)\}$ in the limit of large $\nu$). Hence, for the contributing region we may take $(x^0 - x^3)x^3 \to 0$. In other words, $x_0^2 - x_3^2 \to 0$. Using the micro-causality requirement that in the commutator $x_0^2 - x_3^2 - x_1^2 - x_2^2 \geqslant 0$, we see that $x^2$ itself must tend to zero. Thus, in the Bjorken limit, the region $x^2 = 0$ in the commutator is the one that contributes. This argument is, however, oversimplified. In general, $f_{\mu\nu}(x)$ is a distribution with a presumably finite number of singularities. All these singularities will contribute in the limit of interest. This is because, for such a distribution $f(x)$ (in one variable, say), the limiting Fourier integral $\lim_{\alpha\to\infty} \int \mathrm{d}x\, \mathrm{e}^{\mathrm{i}\alpha x} f(x)$ has contributions from all regions of $x$ where $f$ varies a great deal within an interval $\Delta x$ of the order of $1/\alpha$. The commutator in $f_{\mu\nu}$ is known to have a singularity on the light-cone. In order to have exclusive light-cone dominance in the Bjorken limit, we must make the assumption that there are not strongly varying parts to the commutator, inside the light-cone, which may contribute in the scaling limit. The assertion that any other singularity in $x^0 - x^3$ coming from inside the light-cone cannot compete with the light-cone singularity is very reasonable. Even if the two singularities are of the same order, the former gets suppressed because of the oscillatory factor associated with it and coming from the exponential. Thus, for example, a $\delta(x^2-a^2)$ singularity ($a^2 > 0$) picks up an effective suppression of $\nu^{-3/4}$ relative to the contribution from $\delta(x^2)$ (Frishman 1972).

*Example of a light-cone singularity*

Consider a free scalar field theory. The current is then given by

$$J(x) = :\phi^\dagger(x)\phi(x): = \phi^\dagger(x)\phi(x) - \langle 0|\phi^\dagger(x)\phi(0)|0\rangle, \tag{5.5}$$

where the second term in the right-hand side of eqn (5.5) is a $c$-number. Consider now the commutator of two currents

$$[J(x), J(y)] = [\phi^\dagger(x)\phi(x), \phi^\dagger(y)\phi(y)]. \tag{5.6}$$

But the commutator of two free scalar fields is

$$[\phi(x), \phi^\dagger(y)] = \mathrm{i}\Delta(x-y),$$

hence eqn (5.6) may be rewritten as

$$[J(x), J(y)] = \mathrm{i}\Delta(x-y)\{\phi^\dagger(x)\phi(y) + \phi^\dagger(y)\phi(x)\},$$

or

$$[J(x), J(y)] = \mathrm{i}\Delta(x-y)\{:\phi^\dagger(x)\phi(y): + :\phi^\dagger(y)\phi(x):\} + \mathrm{i}\Delta(x-y)\Delta_1(x-y). \tag{5.7}$$

In eqn (5.7), the disconnected $c$-number part $\mathrm{i}\Delta(x-y)\Delta_1(x-y)$, with

$$\Delta_1(x-y) = \langle 0|\phi^\dagger(x)\phi(y) + \phi^\dagger(y)\phi(x)|0\rangle$$

is the most singular term on the light cone. Take the standard Fourier representations for these functions (Bjorken and Drell 1965),

$$\Delta(x-y) = -\frac{\mathrm{i}}{(2\pi)^3}\int \mathrm{d}^4p\, \mathrm{e}^{-\mathrm{i}p.x}\,\varepsilon(p^0)\delta(p^2-m^2),$$

$$\Delta_1(x-y) = -\frac{\mathrm{i}}{(2\pi)^3}\int \mathrm{d}^4p\, \mathrm{e}^{-\mathrm{i}p.x}\delta(p^2-m^2).$$

These specify the nature of the light-cone singularities contained in the said functions. Thus

$$\lim_{x^2\to 0}\Delta(x) = -\frac{1}{2\pi}\varepsilon(x^0)\delta(x^2) + \text{less-singular terms}, \tag{5.8}$$

$$\lim_{x^2\to 0}\Delta_1(x) = \varepsilon(x^0)\delta'(x^2) + \text{less-singular terms}. \tag{5.9}$$

Eqns (5.7), (5.8), and (5.9) demonstrate explicitly the light-cone singularities in the current commutator of this theory. The property that the disconnected

$c$-number part is more singular on the light-cone than the connected piece, seen in this illustration, is in fact a more general characteristic of commutators and of products of currents. We can also take a free spin-$\frac{1}{2}$ field theory with the current $J_\mu(x) = :\bar{\psi}(x)\gamma_\mu\psi(x):$, $\psi$ being the spinor field, and discuss the light-cone singularities of the commutator $[J_\mu(x), J_\nu(y)]$ (see Exercise 5.1); similar features emerge.

## Light-cone expansions

*OPE near the light-cone*

We start with the scale invariant short-distance OPE (see Part II, p. 88) of Wilson for two local fields $A$ and $B$, as given in eqn (4.32),

$$\lim_{x-y\to 0} A(x)B(y) = \sum_n C_n(x-y)O_n(y). \qquad (5.10)$$

In eqn (5.10) the operators $\{O_n\}$ may be chosen at the point $x$ rather than at $y$. The equation then changes to

$$\lim_{x-y\to 0} A(x)B(y) = \sum_n C_n'(x-y)O_n(x), \qquad (5.11)$$

where $C_n'$ and $C_n$ may be related by means of the formula

$$O_n(x) = \sum_{l=0}^{\infty} \frac{1}{l!}[(x-y).\partial]^l O_n(y).$$

The right-hand side of eqn (5.11) is thus just a rearranged version of the right-hand side of eqn (5.10). Note that any $c$-number tensorial function, regular on the light-cone, is regular at short distances (i.e. at the tip of the light-cone), but the converse is not true. Thus $(x-y)_\alpha(x-y)_\beta/(x-y)^2$ is singular on the light-cone but is regular in the limit $\rho \to 0$ when $(x-y)_\alpha = \rho\eta_\alpha$ and $\eta_\alpha$ is a fixed four-vector. Therefore it is hard to write scale-invariant OPEs on the light-cone than at short distances; many more types of singularities compatible with scale invariance are possible in the former case as compared with the latter situation. Among other things, as we shall see, on the light-cone an infinite number of terms will be needed to any order of singularity. Now consider a singularity near the light-cone for the commutator $[A(z), B(0)]$ of the type

$$\{(-z^2+i\varepsilon z^0)^{-\lambda} - (-z^2-i\varepsilon z^0)^{-\lambda}\} z^{\mu_1} \ldots z^{\mu_n} O_{\mu_1\ldots\mu_n}(0). \qquad (5.12)$$

In a matrix element of the tensor operator $O_{\mu_1\ldots\mu_n}$ of eqn (5.12), between the states $\langle\alpha|$ and $|\beta\rangle$, the most important part contributing to the leading light-cone term will be proportional to $p_{1,\mu_1} \ldots p_{n,\mu_n}$, where $p_1, \ldots, p_n$ are momenta present in the states $\langle\alpha|$ and $|\beta\rangle$. Thus the leading light-cone singularity in the expansion of eqn (5.12) must go as $(z^2)^{-\lambda}$. In other terms, tensors such as $g_{\mu_i\mu_k}$ will reduce this singularity, and they will be subdominant. On the other hand, the maximal short-distance singularity is $\rho^{-2\lambda+n}$, where we again write $z_\alpha = \rho\eta_\alpha$. Thus, if $n-2\lambda > 0$, the contribution is regular at short distances despite being singular on the light-cone. It is thus clear that a light-cone OPE cannot be derived from Wilson's short-distance OPE without knowing the latter to an infinite degree of accuracy (including the regular terms).

We shall now introduce the hypothesis of light-cone expansions (Frishman 1971a,b; Brandt and Preparta 1971). For two local operators $A(x)$ and $B(y)$, this hypothesis can be stated as follows:

$$A(x)B(y) \underset{(x-y)^2\to 0}{\simeq} \sum_{n=0}^{\infty} D_n(x-y)O_n(y), \tag{5.13}$$

where the $\{O_n(y)\}$ are an appropriate set of local fields and the $c$-number functions $D_n(x-y)$ include such light-cone singularities as

$$\sum_{m=0}^{\infty} a_m\{-(x-y)^2+\mathrm{i}\varepsilon(x^0-y^0)\}^{-p_m}(x-y)^{\mu_1}\dots(x-y)^{\mu_m}O_{\mu_1\dots\mu_m}(y). \tag{5.14}$$

In eqn (5.14) $p_m = d_A+d_B-m-d_{O_m}$ ($d$ standing, as usual, for asymptotic scale dimensions) in consequence of scale invariance on the light-cone. Furthermore, the leading light-cone terms in the right-hand side of eqn (5.13) are presumed to be invariant under the symmetry group $SU(3)\times SU(3)$, as a generalization of Wilson's ideas; indeed, these terms belong to the skeleton theory. We have already remarked that such a light-cone expansion cannot be obtained without further assumptions from the short-distance OPE. This type of expansion is seen to be true in free field theory. It has been shown (Brandt and Preparata 1971) to hold in certain *formal* interacting field theories. The main feature that distinguishes eqn (5.13) from the short-distance OPE of Wilson (see Part II, p. 88) is the presence of an infinite number of leading terms to any order. The latter is in fact a consequence of locality and causality. With a finite number of terms the right-hand side of eqn (5.13) would commute with another local operator $C(z)$ so long as $z-y$ were space-like; but the corresponding left-hand side would not do so if $z-x$ were time-like. For $(x-y)^2 \to 0$ (although not for $x-y\to 0$), it is certainly possible to find values of $z$ for which $(z-y)^2 < 0$, but $(z-x)^2 > 0$. Thus the existence of a finite number of leading terms to any order of singularity in a light-cone OPE would lead to a contradiction. Hence we need an infinite number of local operators on the right-hand side of eqn (5.13). Another way of saying the same thing is that, if we were to collect together all terms of a given type of singularity and write a finite series expansion

$$\sum_{n}^{\text{finite}} D_n(x-y)O_n(x, y), \tag{5.15}$$

the operators $O_n$ would no longer be local (this problem obviously does not arise in the short-distance expansion). We shall see later that it is convenient to introduce non-local operators to make the number of terms on the right-hand sides of light-cone operator-products finite and that such operators have very interesting features.

*Light-cone expansion of two electromagnetic currents and its implication*

We consider the product of two electromagnetic currents (the discussion is easily generalizable to the weak case) and recall from our discussions of short-distance scale invariance (eqn (4.56), p. 96) that

$$J_\mu^{\text{EM}}(z)J_\nu^{\text{EM}}(0) \underset{z\to 0}{\sim} \cdots + \sum_i C^i_{\mu\nu\alpha\beta}(z)\phi_i^{\alpha\beta}(0) + \cdots, \tag{5.16}$$

where $\phi_i^{\alpha\beta}$ are spin-2 fields with $d = 4$ and

$$\begin{aligned} C^i_{\mu\nu\alpha\beta}(z) = {} & B^i_{\text{L}}(-g_{\mu\nu}\Box + \partial_\mu\partial_\nu)\{z_\alpha z_\beta(-z^2+\mathrm{i}\varepsilon z^0)\} + B^i_2(-g_{\mu\alpha}g_{\nu\beta}\Box + \\ & + g_{\mu\alpha}\partial_\nu\partial_\beta + g_{\nu\beta}\partial_\alpha\partial_\mu - 2g_{\mu\nu}\partial_\alpha\partial_\beta - g_{\nu\alpha}g_{\mu\beta}\Box + \\ & + g_{\nu\alpha}\partial_\mu\partial_\beta + g_{\mu\beta}\partial_\alpha\partial_\nu)\tfrac{1}{2}\ln(-z^2+\mathrm{i}\varepsilon z^0). \end{aligned} \tag{5.17}$$

In eqn (5.16) we have displayed only the term that gave the most spectacular contribution to spin-averaged deep inelastic eN scattering in the scaling limit, as discussed in Chapter 4. In analogy with eqn (5.17), the light-cone expansion may be written as

$$\begin{aligned} & [J_\mu^{\text{EM}}(x), J_\nu^{\text{EM}}(y)] \\ & \underset{z^2=(x-y)^2\to 0}{\simeq} \cdots + (\partial^y_\mu\partial^x_\nu - \partial^x_\lambda\partial^\lambda_y g_{\mu\nu})[\{(-z^2+\mathrm{i}\varepsilon z^0)^{-1} - \\ & \qquad -(-z^2-\mathrm{i}\varepsilon z^0)^{-1}\}U(x, y)] + \\ & \qquad +(-\partial^x_\lambda\partial^\lambda_y g_{\mu\alpha}g_{\nu\beta} + g_{\mu\alpha}\partial^y_\beta\partial^x_\nu + g_{\nu\beta}\partial^x_\alpha\partial^y_\mu - g_{\mu\nu}\partial^x_\alpha\partial^y_\beta - \\ & \qquad -\partial^y_\lambda\partial^\lambda_x g_{\nu\alpha}g_{\mu\beta} + g_{\nu\alpha}\partial^x_\beta\partial^y_\mu + g_{\mu\beta}\partial^y_\alpha\partial^x_\nu - g_{\mu\nu}\partial^y_\alpha\partial^x_\beta) \times \\ & \qquad \times \tfrac{1}{2}[\{\ln(-z^2+\mathrm{i}\varepsilon z^0) - \ln(-z^2-\mathrm{i}\varepsilon z^0)\} \times \\ & \qquad \times V^{\alpha\beta}(x, y)] + \cdots. \end{aligned} \tag{5.18}$$

In eqn (5.18) $U(x, y)$ and $V^{\alpha\beta}(x, y)$ are non-local operators which may be expressed as infinite series of local operators. In these series, each term in $U$ and $V^{\alpha\beta}$ must have asymptotic scale dimensions of 2 and 4 respectively.

Thus we have again satisfied the canonical rule $d_J = J+2$ to which we alluded in Chapter 4. Scale invariance is preserved so long as this relation is maintained. In the short-distance limit when $(x-y)_\mu \to 0$, we must have

$$\lim_{x-y\to 0} U(x, y) = \sum_i B^i_{\mathrm{L}} z_\alpha z_\beta \phi_i^{\alpha\beta}(y),$$

$$\lim_{x-y\to 0} V^{\alpha\beta}(x, y) = \sum_i B^i_2 \phi_i^{\alpha\beta}(y). \tag{5.19}$$

Eqns (5.19) connect the light-cone expansion (eqn (5.18)) with the short-distance expansion (eqn (5.17)). Antisymmetry under the changes $\mu \leftrightarrow \nu$ and $x \leftrightarrow y$, manifest in the commutator, implies

$$U(x, y) = U(y, x) \quad \text{and} \quad V^{\alpha\beta}(x, y) = V^{\alpha\beta}(y, x).$$

Therefore the expansions of these non-local objects into local terms may be written as

$$U(z, 0) = U(0) + \frac{z_\alpha z_\beta}{2!} U^{\alpha\beta}(0) + \cdots,$$

$$V^{\alpha\beta}(z, 0) = V^{\alpha\beta}(0) + \frac{z_\gamma z_\delta}{2!} V^{\alpha\beta\gamma\delta}(0) + \cdots. \tag{5.20}$$

The coefficients of the powers of $z$ in eqns (5.20) are local operators obeying the rule $d_J = J+2$.

We shall now show how eqn (5.18), along with the $d_J = J+2$ rule, immediately leads to Bjorken scaling in deep inelastic eN scattering. This will be complementary to the treatment given in Chapter 4. Let us write the spin-averaged structure function as

$$(2\pi)^2 \frac{p_0}{M} \int \mathrm{d}^4 x\, \mathrm{e}^{\mathrm{i}q.x} \langle \mathrm{N}|[J^{\mathrm{EM}}_\mu(x), J^{\mathrm{EM}}_\nu(0)]|\mathrm{N}\rangle$$

$$\equiv W^{\mathrm{e}}_{\mu\nu}$$

$$= W^{\mathrm{e}}_{\mathrm{L}}\left(-g_{\mu\nu} + \frac{q_\mu q_\nu}{q^2}\right) + \frac{W^{\mathrm{e}}_2}{M^2 q^2} \times$$

$$\times \{q^2 p_\mu p_\nu - q \,.\, p(q_\mu p_\nu + q_\nu p_\mu) + (q \,.\, p)^2 g_{\mu\nu}\}. \tag{5.21}$$

In eqn (5.21) $|\mathrm{N}\rangle$, as usual, is a nucleon state of four-momentum $p$. In the light-cone limit we have to consider the matrix elements of $U$, $V^{\alpha\beta}$, ..., etc. between the states $\langle \mathrm{N}|$ and $|\mathrm{N}\rangle$. The leading terms will invariably be of the form $(2\pi)^{-3} M p_0^{-1}(b_0, b_2 p^\alpha p^\beta, \ldots)$ (terms involving $g_{\alpha\beta}$, etc. will give non-leading contributions upon contraction with $x^\alpha x^\beta$, etc.), where the $b$s are dimensionless constants. Now define

$$g(p \,.\, x) \equiv b_0 + \frac{1}{2!} b_2 (x \,.\, p)^2 + \frac{1}{4!} b_4 (x \,.\, p)^4 + \cdots. \tag{5.22}$$

Then, eqns (5.18), (5.21), and (5.22) imply that

$$\lim_{\mathrm{Bj}} W_{\mathrm{L}}^{\mathrm{e}} = (-q^2)\frac{\mathrm{i}}{2\pi}\int \mathrm{d}^4x\, \mathrm{e}^{\mathrm{i}q.x} g(x\,.\,p)\delta(x^2)\varepsilon(x^0). \tag{5.23}$$

In obtaining eqn (5.23), we have used the fact that

$$\mathrm{Im}(-z^2+\mathrm{i}\varepsilon z^0)^{-1} = -\pi\delta(z^2)\varepsilon(z^0).$$

Defining the Fourier transform of $g(x\,.\,p)$ as

$$G(\xi) \equiv \frac{1}{2\pi}\int_{-\infty}^{\infty} \mathrm{d}(x\,.\,p) g(x\,.\,p)\,\mathrm{e}^{-\mathrm{i}\xi x.p}, \tag{5.24}$$

eqn (5.23) may now be rewritten as

$$\lim_{\mathrm{Bj}} W_{\mathrm{L}}^{\mathrm{e}} = \mathrm{i}q^2 \int_{-\infty}^{\infty} \mathrm{d}\xi \int \mathrm{d}^4x\, \mathrm{e}^{\mathrm{i}(q+\xi p).x}\delta(x^2)\varepsilon(x^0)G(\xi). \tag{5.25}$$

On using the relation

$$\int \mathrm{d}^4z\, \mathrm{e}^{\mathrm{i}R.z}\delta(z^2) = -\frac{4\pi^2}{\mathrm{i}}\delta(R^2)\varepsilon(R^0), \tag{5.26}$$

eqn (5.25) becomes

$$\begin{aligned}\lim_{\mathrm{Bj}} W_{\mathrm{L}}^{\mathrm{e}} &= -q^2(2\pi)^2 \int_{-\infty}^{\infty} \mathrm{d}\xi\, \varepsilon(q^0+\xi p^0)\delta(q^2+2\xi p\,.\,q+\xi^2M^2)G(\xi) \\ &\simeq (2\pi)^2 w \int_{-\infty}^{\infty} \mathrm{d}\xi\, \delta(\xi-w)G(\xi) \\ &= (2\pi)^2 wG(w). \end{aligned} \tag{5.27}$$

Thus, the structure function $W_{\mathrm{L}}^{\mathrm{e}}$ scales in the Bjorken limit, since there is no reason why $G(w)$ should not exist. Similarly, the leading tensors of $V^{\alpha\beta}$, $V^{\alpha\beta\gamma\delta}$, ... , etc. between $\langle \mathrm{N}|$ and $|\mathrm{N}\rangle$ can be written as $(2\pi)^{-3}Mp_0^{-1}(C_2p^\alpha p^\beta$, $C_4p^\alpha p^\beta p^\gamma p^\delta, \ldots)$, where the $C$s are dimensionless. We define

$$h(x\,.\,p) \equiv C_2+\frac{C_4}{2!}(x\,.\,p)^2+\frac{C_4}{4!}(x\,.\,p)^4+\cdots \tag{5.28}$$

and use the relation $\mathrm{Im}\{\ln(-z^2+\mathrm{i}\varepsilon z^0)\} = -\pi\theta(z^2)\varepsilon(z^0)$ to write

$$\lim_{\mathrm{Bj}} W_2^{\mathrm{e}} = M^2\frac{q^2}{2\pi}\int \mathrm{d}^4x\, \mathrm{e}^{\mathrm{i}q.x}(-2\pi\mathrm{i})\theta(x^2)\varepsilon(x^0)h(x\,.\,p). \tag{5.29}$$

We may now introduce the Fourier transform

$$H(\xi) = \frac{1}{2\pi} \int_{-\infty}^{\infty} \mathrm{d}(x\,.\,p)\,\mathrm{e}^{-\mathrm{i}\xi x.p} h(x\,.\,p)$$

and, proceeding as with $W_{\mathrm{L}}$, show (see Exercise 5.1) that

$$\lim_{\mathrm{Bj}} \nu W_2^{\mathrm{e}} = 4\pi M w \int_{-\infty}^{\infty} \mathrm{d}\xi\, \delta(\xi - w) H(\xi)$$

$$= 4\pi M w H(w). \tag{5.30}$$

Thus $\nu W_2^{\mathrm{e}}$ also scales in the deep inelastic limit.

## Light-cone current algebra

### *Motivation and structure*

It has been shown (Fritzsch and Gell-Mann 1971) that the light-cone expansions of Frishman, Brandt and Preparata can be cast into a beautiful and compact mathematical form capable of generalization to a closed Lie algebra. This form is an extension of the U(6) × U(6) equal-time current algebra (Adler and Dashen 1968) to the light-cone. For that purpose, we have to consider certain non-local operators $O(x, y)$, such as the $U(x, y)$ and $V^{\alpha\beta}(x, y)$ introduced earlier, and investigate their properties. We know that these operators are of a very special type. They are defined as non-singular on the light cone $(x-y)^2 = 0$ and represent infinite sums of local operators all satisfying the rule $d_J = J+2$. Such an object is called a bilocal operator. The advantage of introducing a bilocal operator is that we have to consider only a finite number of these in a light-cone expansion, as opposed to an infinite series of local operators. Furthermore, these bilocal operators have definite and very nice symmetry properties and may, indeed, themselves satisfy certain closed algebras. Before jumping very far, however, we want to see what we gain in terms of compactness and convenience just by introducing these new operators. To this end, we can make abstractions from the free field theory of quarks (Gell-Mann and Nee'man 1964). Such abstractions are well known to have been eminently successful for equal-time current algebra. Moreover, as will be shown shortly, the leading light-cone singularities of the commutator of two hadronic electromagnetic currents are known from the SLAC experiments on deep inelastic eN scattering to be the same as those of a massless free quark-field theory. A similar (though less direct) statement can be made for the commutator of two hadronic weak currents, based on the existing data on inelastic $(\nu, \bar{\nu})$ scattering from nucleons. First therefore we shall accept the hypothesis that the leading terms in the light-cone current commutators are those of the free massless quark-field theory—unaffected by the strong interactions present. Later on, we shall view this question formally in the presence of an interaction mediated by a vector gluon.

We now discuss light-cone commutators based on the above hypothesis. Taking the local hadronic vector and axial vector currents

$$J^i_\mu(x) \sim \,:\bar{q}(x)\gamma_\mu \frac{\lambda^i}{2} q(x):, \tag{5.31a}$$

$$J^i_{\mu 5}(x) \sim \,:\bar{q}(x)\gamma_\mu \gamma_5 \frac{\lambda^i}{2} q(x):, \tag{5.31b}$$

we define the bilocal currents in terms of the quark constituents as follows:

$$J^i_\mu(x, y) \sim \,:\bar{q}(x)\gamma_\mu \frac{\lambda^i}{2} q(y):, \tag{5.32a}$$

$$J^i_{\mu 5}(x, y) \sim \,:\bar{q}(x)\gamma_\mu \gamma_5 \frac{\lambda^i}{2} q(y):. \tag{5.32b}$$

Moreover, we introduce the symmetric and antisymmetric combinations

$$J_\mu^{\oplus/\ominus, i}(x, y) \sim \tfrac{1}{2}\!:\!\left\{\bar{q}(x)\gamma_\mu \frac{\lambda^i}{2} q(y) \pm \bar{q}(y)\gamma_\mu \frac{\lambda^i}{2} q(x)\right\}\!: \tag{5.33}$$

and similarly for the axial vector currents. For free massless quark fields we have

$$\{q(x), \bar{q}(y)\} = \mathrm{i}S(x-y) = \gamma\,.\,\partial\Delta(x-y), \tag{5.34}$$

with (cf. eqn (5.8))

$$\lim_{z^2\to 0} \Delta(z) = -\frac{1}{2\pi}\varepsilon(z^0)\delta(z^2). \tag{5.35}$$

Consider the commutator

$$[J^i_\mu(x), J^j_\nu(y)] \sim \left[\bar{q}(x)\gamma_\mu \frac{\lambda^i}{2} q(x), \bar{q}(y)\gamma_\nu \frac{\lambda^j}{2} q(y)\right]$$

near $(x-y)^2 = 0$. (It is convenient to use a decomposition of the type

$$[A\alpha B, C\beta D] = A\alpha\{B, C\}\beta D - A\alpha C\{B, BD\} + \{A, \alpha C\}\beta DB - C\beta\{A, \alpha D\}B,$$

where $A, \alpha, B, C, \beta, D$ are arbitrary operators.) Then we have

$$[J^i_\mu(x), J^j_\nu(y)] \underset{(x-y)^2\to 0}{\sim} \left\{\bar{q}(x)\frac{\lambda^i\lambda^j}{4}\gamma_\mu\gamma_\alpha\gamma_\nu q(y) + \bar{q}(y)\frac{\lambda^j\lambda^i}{4}\gamma_\nu\gamma_\alpha\gamma_\mu q(x)\right\}\{-\partial^\alpha_z\,\Delta(z)\},$$

where $z = x-y$. On using the relations

$$\lambda^i\lambda^j = \mathrm{i}f^{ijk}\lambda^k + d^{ijk}\lambda^k$$

and

$$\gamma_\mu\gamma_\alpha\gamma_\nu = (g_{\mu\alpha}g_{\nu\sigma} + g_{\nu\alpha}g_{\mu\sigma} - g_{\mu\nu}g_{\alpha\sigma})\gamma^\sigma - \mathrm{i}\varepsilon_{\mu\nu\alpha\sigma}\gamma^\sigma\gamma_5,$$

and after some algebraic manipulations, we obtain the following results:

$$\begin{aligned}
&[J^i_\mu(x), J^j_\nu(y)] \\
&\quad \hat{=} [J^i_{\mu 5}(x), J^j_{\nu 5}(y)] \\
&\quad \hat{=} \frac{1}{2\pi}\partial^z_\rho(\varepsilon(z^0)\delta(z^2))[\mathrm{i}f^{ijk}\{S_\mu{}^\rho{}_\nu{}^\sigma J^{\oplus,k}_\sigma(x,y) + \mathrm{i}\varepsilon_\mu{}^\rho{}_\nu{}^\sigma J^{\ominus,k}_{\sigma 5}\} + \\
&\qquad + d^{ijk}\{S_\mu{}^\rho{}_\nu{}^\sigma J^{\ominus,k}_\sigma(x,y) + \mathrm{i}\varepsilon_\mu{}^\rho{}_\nu{}^\sigma J^{\oplus,k}_{\sigma 5}\}],
\end{aligned} \tag{5.36}$$

$$\begin{aligned}
&[J^i_{\mu 5}(x), J^j_\nu(y)] \\
&\quad \hat{=} [J^i_\mu(x), J^j_{\nu 5}(y)] \\
&\quad \hat{=} \frac{1}{2\pi}\partial^z_\rho\{\varepsilon(z^0)\delta(z^2)[\mathrm{i}f^{ijk}\{S_\mu{}^\rho{}_\nu{}^\sigma J^{\oplus,k}_{\sigma 5}(x,y) + \mathrm{i}\varepsilon_\mu{}^\rho{}_\nu{}^\sigma J^{\ominus,k}_\sigma(x,y)\} + \\
&\qquad + d^{ijk}\{S_\mu{}^\rho{}_\nu{}^\sigma J^{\ominus,k}_{\sigma 5}(x,y) + \mathrm{i}\varepsilon_\mu{}^\rho{}_\nu{}^\sigma J^{\oplus,k}_\sigma(x,y)\}].
\end{aligned} \tag{5.37}$$

In eqns (5.36) and (5.37)

$$S_\mu{}^\rho{}_\nu{}^\sigma = \delta^\rho_\mu\delta^\sigma_\nu + \delta^\sigma_\mu\delta^\rho_\nu - g_{\mu\nu}g^{\rho\sigma},$$

and the symbol $\hat{=}$ means the leading connected light-cone singularity in the left-hand side. The commutators of the two equations (5.36) and (5.37) constitute the light-cone current algebra of Fritzsch and Gell-Mann. Note that we have displayed only the connected parts of the commutators and ignored the disconnected contributions. The connected parts are what will interest us here, but we shall return briefly to the disconnected parts later.

*Current conservation*

We have to ensure that the leading light-cone behaviour of current commutators as discussed above is consistent with current conservation (Gross and Treiman 1971). We recall from the discussion given earlier that the leading light-cone terms are invariant under $SU(3) \times SU(3)$ symmetry because the structure of these terms is abstracted from the free massless quark model. Thus all the $SU(3) \times SU(3)$ currents have to be conserved to leading order on the light-cone. Eqns (5.36) and (5.37) will be shown now to be consistent with current conservation, by the explicit use of the quark structure of the local and bilocal currents. To keep the discussion simple, we shall refer specifically to the free massless quark-field theory. This, however, may be circumvented, and the same result established in general (albeit only to leading order on the light-cone) by a more involved argument (Frishman 1972) (see also Exercise 5.2). For our purpose it is convenient to define $X = \frac{1}{2}(x+y)$ and $u \equiv \frac{1}{2}(x-y) = \frac{1}{2}z$. We take eqn (5.36) (the entire analysis goes through for eqn (5.37) in an identical manner) and rewrite it as

$$[J^i_\mu(x), J^j_\nu(y)] \triangleq -\Theta^{ij}_{\mu\rho\nu}(X, u)\partial^\rho_u \, \Delta(u), \tag{5.38}$$

where $\Delta(u)$ is as defined in eqn (5.8) and

$$\Theta^{ij}_{\mu\rho\nu} = \mathrm{i}f^{ijk}\{S_{\mu\rho\nu}{}^\sigma J^{\oplus,k}_\sigma(x, y) + \mathrm{i}\varepsilon_{\mu\rho\nu}{}^\sigma J^{\ominus,k}_{\sigma 5}(x, y)\} + \\ + d^{ijk}\{S_{\mu\rho\nu}{}^\sigma J^{\ominus,k}_\sigma(x, y) + \mathrm{i}\varepsilon_{\mu\rho\nu}{}^\sigma J^{\oplus,k}_{\sigma 5}(x, y)\}. \tag{5.39}$$

We now use the result that, near the light-cone, $\Box_u \Delta(u) = -(1/2\pi)\Box_u \times \times \{\varepsilon(u^0)\delta(u^2)\}$ to write

$$\Theta^{ij}_{\mu\rho\nu}\partial^\mu_u\partial^\rho_u\Delta(u) = 0 = \Theta^{ij}_{\mu\rho\nu}\partial^\nu_u\partial^\rho_u\Delta(u). \tag{5.40}$$

Eqn (5.40) follows because the part of the tensor $\Theta_{\mu\rho\nu}$ that is symmetric under the interchange of $\mu$ and $\rho$ is proportional to the metric tensor $g_{\mu\rho}$. Because of eqns (5.38) and (5.40), the statement of current conservation can be construed as

$$\frac{\partial}{\partial x_\mu}\Theta^{ij}_{\mu\rho\nu} \triangleq 0 \triangleq \frac{\partial}{\partial y_\nu}\Theta^{ij}_{\mu\rho\nu}. \tag{5.41}$$

Now the results

$$\frac{\partial}{\partial X_\sigma}J^i_\sigma(x, y) \triangleq 0 \triangleq \frac{\partial}{\partial u_\sigma}J^i_\sigma(x, y)$$

which follow trivially in the free-quark model with $J^i_\sigma(x, y) \sim :\bar{q}(x)\gamma_\sigma(\lambda^i/2)q(x):$ simplify the divergence of eqn (5.41) to

$$\begin{aligned} \frac{\partial}{\partial x_\mu}\Theta^{ij}_{\mu\rho\nu} &= \frac{1}{2}\left(\frac{\partial}{\partial X_\mu} + \frac{\partial}{\partial x_\mu}\right)[\mathrm{i}f^{ijk}\{(g_{\mu\rho}\delta^\sigma_\nu - g_{\mu\nu}\delta^\rho_\sigma)J^{\oplus,k}_\sigma(x, y) + \\ &\quad + \mathrm{i}\varepsilon_{\mu\rho\nu}{}^\sigma J^{\ominus,k}_{\sigma 5}(x, y)\} + \mathrm{d}^{ijk}\{(g_{\mu\rho}\delta_\sigma - g_{\mu\nu}\delta^\sigma_\rho)J^{\ominus,k}_\sigma(x, y) + \\ &\quad + \mathrm{i}\varepsilon_{\mu\rho\nu}{}^\sigma J^{\oplus,k}_{\sigma 5}(x, y)\}] \\ &= \tfrac{1}{2}[\mathrm{i}f^{ijk}\{\hat{\partial}_\rho J^{\oplus,k}_\nu(x, y) - \hat{\partial}_\nu J^{\oplus,k}_\rho(x, y) + \mathrm{i}\varepsilon_{\mu\rho\nu}{}^\sigma\partial^\mu J^{\ominus,k}_{\sigma 5}(x, y)\} + \\ &\quad + \mathrm{d}^{ijk}\{\hat{\partial}_\rho J^{\ominus,k}_\nu(x, y) - \hat{\partial}_\nu J^{\ominus,k}_\rho(x, y) + \mathrm{i}\varepsilon_{\mu\rho\nu}{}^\sigma\hat{\partial}^\mu J^{\oplus}_{\sigma 5}(x, y)\}. \end{aligned} \tag{5.42}$$

In eqn (5.42) $\hat{\partial} \equiv \partial/\partial X + \partial/\partial u$. Further we may write

$$\hat{\partial}_\rho J^{\overset{\oplus}{\ominus},k}_\nu(x, y) - \hat{\partial}_\nu J^{\overset{\oplus}{\ominus},k}_\rho(x, y) \triangleq -\mathrm{i}\varepsilon_{\mu\rho\nu}{}^\sigma\hat{\partial}^\mu J^{\overset{\ominus}{\oplus},k}_{\sigma 5}(x, y) \tag{5.43}$$

Eqn (5.43) follows trivially in the free massless quark-field theory, because of the relation

$$(\partial_\rho\gamma_\nu - \partial_\nu\gamma_\rho)q = -\mathrm{i}\varepsilon_{\mu\rho\nu}{}^{\sigma}\partial^\mu\gamma_\sigma\gamma_5 q$$

in that theory. However, eqns (5.43) and (5.42) lead to the vanishing of the left-hand side of the latter and hence of $(\partial/\partial y_\nu)\Theta^{ij}_{\mu\rho\nu}$. Thus eqn (5.41) and hence current conservation, is established.

*Bilocal algebra*

Eqns (5.36) and (5.37) relate commutators of local currents on the light-cone to bilocal currents. They provide us with a compact and elegant way of handling the original light-cone operator product expansions. However, they do not form a closed algebra by themselves, and are in that sense incomplete. In order to have a closed algebra we must have commutation relations among bilocal currents only. These are identified by their quark structure:

$$J^i_\mu(x, y) \sim \,:\bar{q}(x)\gamma_\mu\frac{\lambda^i}{2}q(y):,$$

$$J^i_{\mu 5}(x, y) \sim \,:\bar{q}(x)\gamma_\mu\gamma_5\frac{\lambda^i}{2}q(y):.$$

If we take their commutators and employ the free quark-field commutation relation (eqn (5.34)), we have

$$\begin{aligned}
&[J^i_\mu(x, u), J^j_\nu(y, v)]\\
&\quad \mathrel{\widehat{=}} [J^i_{\mu 5}(x, u), J^j_{\nu 5}(y, v)]\\
&\quad \mathrel{\widehat{=}} -\partial^\rho\Delta(x-v)(\mathrm{i}f^{ijk}-d^{ijk})\{S_{\mu\rho\nu\sigma}J^{\sigma,k}(y, u)-\mathrm{i}\varepsilon_{\mu\rho\nu\sigma}J^{\sigma,k}_5(y, u)\}-\\
&\qquad -\partial^\rho\Delta(u-y)(\mathrm{i}f^{ijk}+d^{ijk})\{S_{\mu\rho\nu\sigma}J^{\sigma,k}(x, v)+\mathrm{i}\varepsilon_{\mu\rho\nu\sigma}J^{\sigma,k}_5(x, v)\},
\end{aligned} \tag{5.44}$$

$$\begin{aligned}
&[J^i_{\mu 5}(x, u), J^j_\nu(y, v)]\\
&\quad \mathrel{\widehat{=}} [J^i_\mu(x, u), J^j_{\nu 5}(y, v)]\\
&\quad \mathrel{\widehat{=}} -\partial^\rho\Delta(x-v)(\mathrm{i}f^{ijk}-d^{ijk})\{S_{\mu\rho\nu\sigma}J^{\sigma,k}_5(y, u)-\mathrm{i}\varepsilon_{\mu\rho\nu\sigma}J^{\sigma,k}(y, u)\}-\\
&\qquad -\partial^\rho\Delta(u-y)(\mathrm{i}f^{ijk}+d^{ijk})\{S_{\mu\rho\nu\sigma}J^{\sigma,k}_5(x, v)+\mathrm{i}\varepsilon_{\mu\rho\nu\sigma}J^{\sigma,k}(x, v)\}.
\end{aligned} \tag{5.45}$$

In eqns (5.44) and (5.45), we have—as usual—

$$\Delta(z) \mathrel{\widehat{=}} -\frac{1}{2\pi}\varepsilon(z^0)\delta(z^2).$$

Eqns (5.44) and (5.45) may be regarded as a closed Lie algebra. However, we now have four coordinates $x$, $y$, $u$, and $v$ to deal with. What, then, is the correct definition of the light-cone and the meaning of the symbol $\hat{=}$ in the present context? Fritzsch and Gell-Mann guessed that the above algebra would hold when all possible distances were light-like, i.e.

$$(x-y)^2 = (x-u)^2 = (x-v)^2 = (y-u)^2 = (y-v)^2 = (u-v)^2 = 0. \quad (5.46)$$

The problem of defining a bilocal current and of testing the validity of the bilocal algebra has also been investigated formally (Gross and Treiman 1971) in a canonical quark–gluon field theory. This approach is formal in that manipulations with the equations of motion are made instead of renormalized perturbation calculations. Gross and Treiman have found that eqns (5.44) and (5.45) are valid and consistent provided we make the identification

$$J^i_\mu(x, y) \underset{(x-y)^2 \to 0}{\sim} :\bar{q}(x)\gamma_\mu \frac{\lambda^i}{2} \exp\left\{ i \int_x^y dz^\mu \, B_\mu(z) \right\} q(y):, \quad (5.47)$$

where $dz^\mu$ is an infinitesimal light-like four-vector element. However, this validity is restricted only to the region where $x$, $y$, $u$, and $v$ are collinear on a light-like ray. This is discussed in detail in what follows next. Let us merely point out here that this collinearity condition is precisely the same as eqn (5.46).

*Bilocal currents and gluon dynamics*

We now consider the formal structure of current commutators in the quark–gluon model which is prescribed by the interaction

$$\mathscr{L}_I = g\bar{q}(x)\gamma_\mu q(x) B^\mu, \quad (5.48)$$

$B_\mu$ being the SU(3)-singlet vector-gluon field. There are other versions of the gluon model with scalar or pseudoscalar gluons. Here we will only consider the interaction of eqn (5.48) for the sake of definiteness, although our results hold (Gross and Treiman 1971) even in the presence of the other types of gluons. We re-emphasize the formal nature of our arguments, which will involve canonical commutators and naive equations of motion rather than renormalized perturbative calculations where (Christ 1972) powerwise violations of leading free field behaviour on the light-cone have been found. (The reader who is not interested in such formal details may skip the present section without loss of continuity.)

The basis of our considerations is the causal quark propagator

$$S_F(x-y) = -i\langle 0|Tq(x)\bar{q}(y)|0\rangle. \quad (5.49)$$

We shall also be interested in the anticommutator

$$S(x, y) = \{q(x), \bar{q}(y)\}. \quad (5.50)$$

Our aim is to rewrite the leading terms of eqns (5.49) and (5.50) on the light-cone $(x-y)^2 = 0$ in a way that simply relates them to the corresponding free-field functions $S_F^{(0)}(x, y)$ and $S^{(0)}(x, y)$. In so far as the leading light-cone singularities of these functions are concerned, it is sufficient to treat $B_\mu$ as an external $c$-number field. This is because, in the following discussion, problems with the ordering of gluon fields at different points will arise only through the commutator

$$[(x-y)^\mu B_\mu(x), (x-y)^\nu B_\nu(y)] \triangleq -\mathrm{i}g_{\mu\nu}(x-y)^2\,\Delta(x-y);$$

this vanishes as $x-y$ approaches a light-like form.

The causal propagator of eqn (5.49) satisfies the equation of motion

$$\mathrm{i}\gamma^\mu\{\partial_\mu^x + \mathrm{i}gB_\mu(x)\}S_F(x-y) = \delta^{(4)}(x-y). \tag{5.51}$$

We now expand

$$S_F(x, y) = \sum_i S_F^{(i)}(x-y) \tag{5.52}$$

as a power series in the coupling constant $g$. Thus we have

$$\begin{aligned} \mathrm{i}\gamma^\mu\partial_\mu^x S_F^{(0)}(x-y) &= \delta^{(4)}(x-y), \\ \mathrm{i}\gamma^\mu\partial_\mu^x S_F^{(1)}(x-y) &= g\not{B}(x)S_F^{(0)}(x-y), \end{aligned} \tag{5.53}$$

etc. Eqns (5.53) may be inverted to

$$\begin{aligned} S_F^{(1)}(x-y) &= g\int \mathrm{d}^4z_1\; S_F^{(0)}(x-z_1)\not{B}(z_1)S_F^{(0)}(z_1-y) \\ &\;\vdots \\ S_F^{(n)}(x-y) &= g^n\int \mathrm{d}^4z_1 \ldots \int \mathrm{d}^4z_n\; S_F^{(0)}(x-z_1)\not{B}(z_1)S_F^{(0)}(z_1-z_2)\ldots \times \\ &\qquad \times \not{B}(z_n)S_F^{(0)}(z_n-y), \end{aligned} \tag{5.54}$$

etc.

In eqns (5.53) and (5.54), $S_F^{(0)}$ is the free-field causal propagator with the leading light-cone term given (ignoring the quark mass) by

$$S_F^{(0)}(x-y) \triangleq \int \frac{\mathrm{d}^4p}{(2\pi)^4}\frac{\not{p}}{p^2+\mathrm{i}\varepsilon}\,\mathrm{e}^{-\mathrm{i}p.(x-y)}. \tag{5.55}$$

We now introduce the variables $X \equiv \frac{1}{2}(x+y)$, $u \equiv \frac{1}{2}(x-y)$ once again and define the Fourier transform

$$B_\mu(x) = \int \mathrm{d}^4q\,\mathrm{e}^{-\mathrm{i}q.x}\tilde{B}_\mu(q). \tag{5.56}$$

The last of eqns (5.54) may now be rewritten as

$$S_{\mathrm{F}}^{(n)}(X,u) \triangleq \frac{g^n}{(2\pi)^4}\prod_{i=1}^{n}\int \mathrm{d}^4 z_i \int \frac{\mathrm{d}^4 p}{p^2+\mathrm{i}\varepsilon}\not{p}\exp\{-\mathrm{i}p\,.\,(x-z_1)\}\times$$

$$\times\int \mathrm{d}^4 q_i \exp(-\mathrm{i}q_i z_i)\tilde{\not{B}}(q_i)\int\frac{\mathrm{d}^4 p_i}{p_i^2+\mathrm{i}\varepsilon}\not{p}_i\exp\{-\mathrm{i}p_i\,.\,(z_i-z_{i+1})\}, \quad (5.57)$$

with $z_{n+1} = y$. Each $z_i$ integration in eqn (5.57) gives rise to a four-dimensional delta function $\delta^{(4)}(p_i+p-\sum_{r=1}^{i} q_r)$, which may be used to dispose of the corresponding $p_i$ integration. Furthermore, if we introduce $P = p-\frac{1}{2}\sum_{i=1}^{n} q_i$, i.e. $p_i = P-\frac{1}{2}\sum_{r=1}^{n} q_r+\sum_{r=i+1}^{n} q_r$ and write

$$p\,.\,X-p_n\,.\,y = 2P\,.\,u+X\,.\sum_{i=1}^{n} q_i,$$

eqn (5.57) becomes

$$S_{\mathrm{F}}^{(n)}(X,u) \triangleq \frac{g^n}{(2\pi)^4}\prod_{i=1}^{n}\int \mathrm{d}^4 q_i\int\frac{\mathrm{d}^4 P}{(P+\frac{1}{2}\sum_{i=1}^{n} q_i)^2+\mathrm{i}\varepsilon}\times$$

$$\times\left(\not{P}+\tfrac{1}{2}\sum_{i=1}^{n}\not{q}_i\right)\exp(-2\mathrm{i}P\,.\,u)\exp\left(-\mathrm{i}X\,.\sum_{r=1}^{n} q_r\right)\times$$

$$\times\frac{\tilde{\not{B}}(q_i)(\not{P}-\frac{1}{2}\sum_{r=1}^{n}\not{q}_r+\sum_{r=i+1}^{n}\not{q}_r)}{(P-\frac{1}{2}\sum_{r=1}^{n} q_r+\sum_{r=i+1}^{n} q_r)^2+\mathrm{i}\varepsilon}. \quad (5.58)$$

Eqn (5.56) and the Riemann–Lebesgue lemma imply—in the manner shown in our previous discussions of light-cone dominance—that the limit when $P_0\to\infty$ with $P^2/P_0$ fixed (i.e. $|\mathbf{P}|\simeq P_0+P^2/2P_0+\ldots$) corresponds to the light-cone. However, in this limit and for the leading terms, we can replace $\not{p}_i$ by $\not{P}$, $p_i^2$ by $P^2+P\,.\,(-\sum_{r=1}^{n} q_r+2\sum_{r=i+1}^{n} q_r)$, and $\not{P}\tilde{\not{B}}\not{P} = 2P\,.\,\tilde{\not{B}}\not{P}-\tilde{\not{B}}P^2$ by $2P\,.\,\tilde{\not{B}}\not{P}$. In other words, we can change $\not{P}\tilde{\not{B}}(q_1)\ldots\tilde{\not{B}}(q_n)\not{P}$ to

$$2^n\not{P}\prod_{i=1}^{n} P\,.\,\tilde{B}(q_n).$$

Then eqn (5.58) takes the form

$$S_{\mathrm{F}}^{(n)}(X,u) \triangleq \frac{2^n g^n}{(2\pi)^4}\int \mathrm{d}^4 P\,\mathrm{e}^{-2iP.u}\prod_{i=1}^{n}\int \mathrm{d}^4 q_i P\,.\,\tilde{B}(q_i)\times$$

$$\times\frac{1}{(P+\frac{1}{2}\sum_{i=1}^{n} q_i)^2+\mathrm{i}\varepsilon}\times$$

$$\times\frac{1}{P^2+P\,.\,(-\sum_{r=1}^{n} q_r+2\sum_{r=i+1}^{n} q_r)}. \quad (5.59)$$

At this stage we 'Feynmanize' the denominators, i.e. write

$$\frac{1}{(P+\frac{1}{2}\sum_{i=1}^{n} q_i)^2+\mathrm{i}\varepsilon}\prod_{i=1}^{n}\frac{1}{P^2+P\,.\,(-\sum_{r=1}^{n} q_r+2\sum_{r=i+1}^{n} q_r)}$$

$$= n!\int_0^1 \mathrm{d}\alpha_1 \ldots \int_0^1 \mathrm{d}\alpha_n\,\delta\left(1-\sum_{i=0}^{n}\alpha_i\right)\frac{1}{\{P^2-P\,.\sum_i q_i(1-2\sum_{j=1}^{i}\alpha_{j-1})\}^{n+1}}. \quad (5.60)$$

If we then substitute $P' = P-\frac{1}{2}\sum_i q_i(1-2\sum_{j=1}^{i}\alpha_{j-1})$ for $P$, the leading light-cone terms in eqn (5.59) may be rewritten as

$$S_{\mathrm{F}}^{(n)}(X,u) \triangleq \frac{2^n g^n}{(2\pi)^4}\int_0^1 \mathrm{d}\alpha_0 \ldots \int_0^1 \mathrm{d}\alpha_n\,\delta\left(1-\sum_{i=0}^{n}\alpha_i\right)\int \mathrm{d}^4P'\,n!\prod_{i=1}^{n}\int \mathrm{d}^4q_i P'\,.\,\tilde{B}(q_i)\times$$

$$\times\exp\left[-\mathrm{i}\sum_{i=1}^{n} q_i\,.\left\{X+u\left(1-\sum_{j=1}^{i}\alpha_{j-1}\right)\right\}\right]\frac{\exp(-2\mathrm{i}P'\,.\,u)}{P'^{2(n+1)}+\mathrm{i}\varepsilon}\not{P}'.$$

(5.61)

We now use the result

$$\int \mathrm{d}^4P'\,P'^{\alpha_1}\ldots P'^{\alpha_n}\not{P}'\frac{\exp(-2\mathrm{i}P'\,.\,u)}{P'^{2(n+1)}+\mathrm{i}\varepsilon} \triangleq \frac{\mathrm{i}^n}{n!}S_{\mathrm{F}}^{(0)}(x-y)u^{\alpha_1}\ldots u^{\alpha_n}$$

in eqn (5.61) and effect a substitution from $\alpha_i$ to $\beta_i \equiv \sum_{j=1}^{i}\alpha_{j-1}$, so that $0 \leqslant \beta_1 \leqslant \beta_2 \ldots \leqslant \beta_n < 1$. This leads to

$$S_{\mathrm{F}}^{(n)}(x,y) = (2\mathrm{i}g)^n S_{\mathrm{F}}^{(0)}(x-y)\prod_{k=1}^{n}\int \mathrm{d}^4q_k u\,.\,\tilde{B}(q_k)\int_0^1 \mathrm{d}\beta_n\int_0^{\beta_n}\mathrm{d}\beta_{n-1}\ldots\int_0^{\beta_2}\mathrm{d}\beta_1\,\times$$

$$\times\exp\left[-\mathrm{i}\sum_{i=1}^{n} q_i\{X+u(1-2\beta_i)\}\right]$$

$$= (2\mathrm{i}g)^n S_{\mathrm{F}}^{(0)}(x-y)\int_0^1 \mathrm{d}\beta_n\int_0^{\beta_n}\mathrm{d}\beta_{n-1}\ldots\int_0^{\beta_2}\mathrm{d}\beta_1\prod_{i=1}^{n}F\{X+u(1-2\beta_i)\},$$

(5.62)

where we have introduced a function $F$ defined as

$$F(z) \equiv \int \mathrm{d}^4q\,\mathrm{e}^{-\mathrm{i}q.z}u\,.\,\tilde{B}(q). \quad (5.63)$$

However, since $F$ is a $c$-number function, we have

$$\int_0^1 \mathrm{d}\beta_n \int_0^{\beta_n} \mathrm{d}\beta_{n-1} \ldots \int_0^{\beta_2} \mathrm{d}\beta_1 F(X, u, \beta_n) \ldots F(X, u, \beta_1)$$

$$= \frac{1}{n!}\int_0^1 \mathrm{d}\beta_n \ldots \int_0^1 \mathrm{d}\beta_1 P\{F(X, u, \beta_n) \ldots F(X, u, \beta_1)\}$$

$$= \frac{1}{n!}\left\{\int_0^1 \mathrm{d}\beta\, F(X, u, \beta\right\}^n, \tag{5.64}$$

where $P\{\ \}$ stands for the Dyson chronological product. With the use of eqns (5.63) and (5.64), eqn (5.62) may be simplified to

$$S_{\mathrm{F}}^{(n)}(x, y) \triangleq \frac{(2\mathrm{i}g)^n}{n!} S_{\mathrm{F}}^{(0)}(x, y)\left\{\int_0^1 \mathrm{d}\beta\, F(X+(1-2\beta)u)\right\}^n.$$

Hence, by eqn (5.52), we see that the light-cone behaviour of the full propagator is

$$\lim_{u^2\to 0} S_{\mathrm{F}}(x, y) = \exp\left\{2\mathrm{i}g \int_0^1 \mathrm{d}\beta\, u\,.\,B(X+(1-2\beta)u)\right\} S_{\mathrm{F}}^{(0)}(x-y),$$

where $B_\mu$ has been transformed back to configuration space via eqn (5.56). Thus

$$\lim_{u^2\to 0} S_{\mathrm{F}}(x, y) = \exp\left(-\mathrm{i}g \int_y^x \mathrm{d}z^\mu\, B_\mu(z)\right) S_{\mathrm{F}}^{(0)}(x-y). \tag{5.65}$$

Similarly we may show that

$$\lim_{u^2\to 0} S(x, y) = \exp\left(-\mathrm{i}g \int_y^x \mathrm{d}z^\mu\, B_\mu(z)\right) S^{(0)}(x, y). \tag{5.66}$$

In eqns (5.65) and (5.66) the path integral in the exponent has to be along a light-like line from $y$ to $x$. If we consider eqns (5.66) and (5.50), it is trivial to see that the bilocal algebra (eqns (5.44) and (5.45)) holds under canonical manipulations of this theory, provided the bilocal current is given the structure of eqn (5.47). Moreover, we can now see the need for the collinearity of the points $x, u, y, v$ (see eqns (5.44) and (5.45)) on a light-like ray. This is because relations such as

$$\exp\left\{-\mathrm{i}g\int_u^x \mathrm{d}z^\mu\, B_\mu(z)\right\}\delta\{(u-y)^2\}\exp\left\{-\mathrm{i}\int_v^y \mathrm{d}z^\mu\, B_\mu(z)\right\}$$

$$= \delta\{(u-y)^2\}\exp\left\{-\mathrm{i}g\int_v^x \mathrm{d}z^\mu\, B_\mu(z)\right\},$$

which play a crucial role in the derivation of the bilocal algebra from eqns (5.47) and (5.50), would not hold otherwise.

*Theoretical basis of free-field behaviour on the light-cone*

Our discussions of light-cone current algebra have centred round one theme: at light-like distances the leading singularities in products of currents are those of free quark-field theory. The implication for an interacting field theory of strong interactions describing the real world is that in certain asymptotic regions it must be free. This will lead to canonical scaling behaviour in those regions; all dimensional anomalies will correspondingly disappear. Thus it is natural to ask: is a field theory with such asymptotic freedom possible? This question may be considered in terms of the renormalization group equation (eqn (4.25)) introduced in Chapter 4. There we have seen that the asymptotic behaviour of renormalized Green's functions is controlled by the zeros of the Gell-Mann–Low function $\beta(\lambda)$. This function is known to have a zero at the origin. Let its derivative at the same point $\beta'(0)$ be positive and let the next zero of the function (for positive values of the bare coupling constant $\lambda$) be at $\lambda_1$. It then follows that in the deep Euclidean domain —as shown in Chapter 4—we obtain scale invariance, but with anomalous scale dimensions evaluated with the value $\lambda_1$ for the renormalized effective coupling constant. On the other hand, suppose $\beta'(0)$ is negative, unlike in the $\lambda\phi^4$ theory considered earlier (in this situation the origin is called an ultraviolet stable point); the behaviour of $\lambda'(t)$ as a function of $t = \frac{1}{2}\ln(s/\mu^2)$ is now completely different and is as shown in Fig. 37. Thus, for $\beta(\lambda') \simeq -|a|\lambda'$ near $\lambda' = 0$, from eqn (4.26) we obtain that $\lambda' \simeq -c\exp(-|a|t)$, $c$ being an arbitrary constant. In other words, the renormalized coupling constant is attenuated to zero at infinite $t$. This theory then is asymptotically free. In such a situation all the canonical properties of free field theory (of free quark-field theory if there is an underlying quark structure) will hold in the scaling limit. Recently there have been several attempts to construct renormalizable models of field theory which have this asymptotic freedom. It has been shown (Zee 1973; Coleman and Gross 1973) that no such theories exist in four dimensions with spinless mesons and spin-$\frac{1}{2}$ fermions only. However, asymptotic freedom may indeed be achieved with spin-1 bosons,

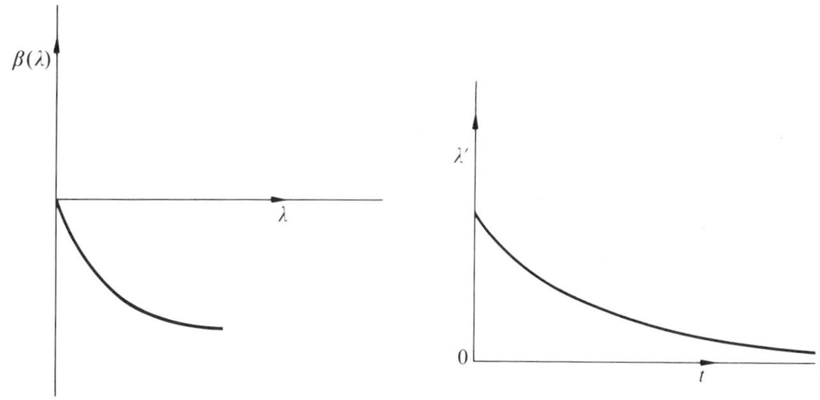


FIG. 37. Asymptotic vanishing of the effective renormalized coupling for an ultraviolet stable origin.

but only if these are gauge fields belonging to a non-Abelian gauge group. Further, there has to be spontaneous symmetry-breaking in order to ensure renormalizability. For the details of such models the reader is referred to the literature (Gross and Wilczek 1973). These do provide a possible insight into the puzzle as to why 'Nature reads textbooks of free field theory on the light-cone'.

**Light-plane formulation**

We shall now discuss an alternative approach (Cornwall and Jackiw 1971) to light-cone physics. This is based on the properties of commutators of currents on the light-plane (also known as the null-plane) $x^+ \equiv (x^0+x^3)/\sqrt{2} \equiv 0$. (By micro-causality, only the edge of contact between this plane and the light-cone contributes to such a commutator.) This formulation of light-cone current algeabra is broadly equivalent to that of Fritzsch and Gell-Mann (Jackiw 1972). However, one noteworthy point of difference in the formalism does exist. In the present scheme a certain subset of the algebra (namely the 'good–good' commutators as explained below) can be obtained by direct appeal to the quantization of the fundamental theory of quarks on the light-plane. No explicit reference to the equations of motion of any formal quark-field theory is necessary for this purpose as opposed to the other case. There is an additional advantage in terms of practical use. It is easier in this approach to tackle certain problems with currents carrying fixed and finite external masses (rather than infinite masses as in deep inelastic reactions); for such situations the relevance of the entire light cone is hard to establish; we shall elaborate further on this in the next chapter. Here we initiate our study of the subject by examining the connections between light-plane commutators and Bjorken scaling and then considering quantization on the light-plane.

*Light-plane current commutators and Bjorken scaling*

It is convenient to use the notations $a^\mu \equiv (a^+, a^r, a^-)$ for the components of any four-vector $a$, where $a^\pm = (a^0 \pm a^3)/\sqrt{2}$ and $r$ stands for the transverse components 1, 2. The components of the metric tensor will be taken as $g^{++} = g^{--} = g^{12} = g^{21} = g^{1\pm} = g^{2\pm} = 0$, $g^{+-} = g^{-+} = g^{11} = g^{22} = 1$. Thus the scalar product of two four-vectors is $a \,.\, b = a^+b^- + a^-b^+ - \mathbf{a}_\perp \,.\, \mathbf{b}_\perp$. Moreover, we use $\partial_- \equiv \partial/\partial x^-$, $\partial_+ \equiv \partial/\partial x^+$. We shall also have occasion to consider the light-like four-vector $n^\mu = (1, 0, 0, -1)/\sqrt{2}$ for which $x \,.\, n = x^+$. The standard variables of deep inelastic $l$N scattering now read $q^2 = 2q^+q^- - q_\perp^2$, $M\nu = p^+q^- + p^-q^+ - \mathbf{p}_\perp \,.\, \mathbf{q}_\perp$. Thus the Bjorken limit is achieved by letting $q^-$ to infinity holding the other components of $q$ fixed. We now return to eqn (5.1). This may be rewritten as

$$W_{\mu\nu}(q) = (2\pi)^2 \frac{p_0}{M} \int_{-\infty}^{\infty} \mathrm{d}x^+ \exp(\mathrm{i}q^-x^+) \int_{-\infty}^{\infty} \mathrm{d}x^- \exp(\mathrm{i}q^+x^-) \times$$

$$\times \int \mathrm{d}^2x_\perp \exp(-\mathrm{i}\mathbf{q}_\perp \,.\, \mathbf{x}_\perp) \langle \mathrm{N}|[J_\mu^\dagger(x), J_\nu(0)]|\mathrm{N}\rangle. \tag{5.67}$$

As $q^- \to \infty$, by the Riemann–Lebesgue theorem, the contribution from the commutator in the integrand of eqn (5.67) is picked out only for the region corresponding to $x^+ \to 0$. Hence the properties of the commutator on the light plane $x^+ = 0$ are quite crucial to the behaviour of the structure tensor in this limit.

We now show how the existence of the 'mildest' (step-discontinuity) type of light-plane singularity in the said commutator leads to a prediction of Bjorken scaling for a certain combination of the $l$N structure functions (Pandit 1973). Define

$$\tilde{W}_{\mu\nu}(x) \equiv (2\pi)^2 \frac{p_0}{M} \langle \mathrm{N}|[J_\mu^\dagger(x), J_\nu(0)]|\mathrm{N}\rangle \tag{5.68}$$

and

$$\mathcal{T}_{\mu\nu}(x^+) \equiv \int \mathrm{d}^4y \, \tilde{W}_{\mu\nu}(y)\delta(n \,.\, y - x^+) f(y), \tag{5.69}$$

where $f(y)$ is a very well-behaved test function. Then we have

$$\mathcal{T}(x^+) \equiv n^\mu n^\nu \mathcal{T}_{\mu\nu}(x^+) = \int \mathrm{d}^4y \, n^\mu n^\nu \tilde{W}_{\mu\nu}(y)\delta(n \,.\, y - x^+) f(y), \tag{5.70}$$

or

$$\mathcal{T}(x^-) = \int \mathrm{d}^2x_\perp \int_{-\infty}^{\infty} \mathrm{d}x^- f(x) \langle \mathrm{N}|[J^{+\dagger}(x), J^+(0)]|\mathrm{N}\rangle. \tag{5.71}$$

Since the commutator in eqn (5.71) is singular on the light-cone $2x^+x^- - x_\perp^2 = 0$, and vanishes outside it, it follows that $\mathscr{T}(x^+)$ must have a singularity at $x^+ = 0$. Consider now the Fourier transform of $\mathscr{T}(x)$, namely,

$$\tilde{\mathscr{T}}(E) \equiv \int \mathrm{d}x^+ \exp(\mathrm{i}Ex^+)\mathscr{T}(x^+) \tag{5.72}$$

in the limits when $E \to \pm\infty$. If $\tau(x^+)$ has only the 'mildest' (step-discontinuity) type of singularity at $x^+ = 0$, it follows from eqn (5.72) that (Lighthill 1964)

$$\lim_{E\to\pm\infty} \tilde{\mathscr{T}}(E) = \frac{\mathrm{i}\{\mathscr{T}(0+) - \mathscr{T}(0-)\}}{E} + O\left(\frac{1}{E^2}\right). \tag{5.73}$$

On the other hand, from eqns (5.70) and (5.72) we have

$$\begin{aligned}\tilde{\mathscr{T}}(E) &= \int \mathrm{d}^4x\, n^\mu n^\nu \tilde{W}_{\mu\nu}(x) f(x)\, \mathrm{e}^{\mathrm{i}En.x} \\ &= \int \mathrm{d}^4k\, n^\mu n^\nu W_{\mu\nu}(k+En)\tilde{f}(-k),\end{aligned} \tag{5.74}$$

where $\tilde{f}(k)$ is the Fourier transform $(2\pi)^{-4}\int \mathrm{d}^4x\, \mathrm{e}^{\mathrm{i}k.x}f(x)$ of the test function. A comparison of eqns (5.73) and (5.74) yields that

$$\lim_{E\to\pm\infty} n^\mu n^\nu W_{\mu\nu}(k+En) = \frac{\Delta(k)}{E} + O\left(\frac{1}{E^2}\right), \tag{5.75}$$

where $\Delta(k)$ is defined by

$$\int \mathrm{d}^4k\, \Delta(k)\tilde{f}(-k) = \mathrm{i}\{\mathscr{T}(0+) - \mathscr{T}(0-)\}$$

and its existence is guaranteed by that of the right-hand side. Substituting $q = k + En$, we can use the information of eqn (5.75) in eqn (5.67). In the limit concerned $q^2 \simeq eEk\,.\,n$ and $\nu \simeq Ep\,.\,n/M$; thus, as $E \to \infty$, we reach the Bjorken limit with the scale variable $w = -k\,.\,n/p\,.\,n$. Comparing the left-hand side of eqn (5.75) with the covariant decompositions of $W_{\mu\nu}$—i.e. eqns (I.7) and (I.27)—in the limit of interest, we obtain

$$\lim_{\mathrm{Bj}} (\nu W_2 - 2wMW_1) = \frac{2}{p\,.\,n}\Delta(k), \tag{5.76}$$

in other words the combination $\nu W_2 - 2wMW_1$ is independent of $E$ (i.e. of $\nu$ or $q^2$) in this limit, and hence scales.

*Light-plane quantization*

The preceding considerations highlight the need for a theory of the singularity structure of current commutators on the light-plane, which we shall now outline. We start with a fundamental quantization procedure on the light-plane. First, define light-like charges $\hat{Q}^i(x^+)$ from the standard hadronic currents $J^i_\mu$ as follows:

$$\hat{Q}^i(x^+) \equiv \int d^4y\, n^\mu J^i_\mu(y)\delta(n\,.\,y - x^+)$$

$$= \int d^2x_\perp \int dx^-\, J^{i,+}(x). \tag{5.77}$$

In analogy with the SU(3) × SU(3) algebra of ordinary charges

$$Q^i(x^0) \equiv \int d^3x\, J^i_0(x)$$

postulated by Gell-Mann (Adler and Dashen 1968), we can advance the hypothesis of a similar algebra obeyed by light-like charges with commutators (Leutwyler 1969) such as

$$[\hat{Q}^i(x^+), \hat{Q}^j(x^+)] = \mathrm{i}f^{ijk}\hat{Q}^k(x^+). \tag{5.78}$$

Eqns (5.77) and (5.78) immediately suggest—again in analogy with the equal-time case—the existence of charge–current and current–current commutation relations such as

$$[\hat{Q}^i(x^+), J^j_\mu(0)]_{x^+=0} = \mathrm{i}f^{ijk}J^k_\mu(0), \tag{5.79a}$$

$$[J^{i,+}(x^+), J^j_\mu(0)]_{x^+=0} = \mathrm{i}f^{ijk}J^k_\mu(0)\delta^{(2)}(\mathbf{x}_\perp)\delta(x^-) + \partial_- S^{ij}_\mu(x) + + \partial_r S^{ij,r}_\mu(x). \tag{5.79b}$$

In the right-hand side of eqn (5.79b) the terms with derivatives vanish on integration over $x^-$ and $\mathbf{x}_\perp$, and are analogous to Schwinger terms. The equal-time commutators of charges and currents could be obtained by formal arguments starting from the fundamental equal-time quantization conditions of an underlying field theory. This suggests that the formal foundation of relations such as eqns (5.79) may be laid in terms of a field theory quantized on the light-plane $x^+ = 0$ rather than on the 'instant' plane $x^0 = 0$. The equivalence between the two quantization schemes is established if the $S$-matrix for either can be shown to be the same. The following plausibility arguments suggest that this is in fact the case (Jackiw 1972a).

1. The corresponding free theories have been shown (Kogut and Soper 1970) to be the same.
2. The hyperplane $x^0 = 0$ is mapped into the light-plane $x^+ = 0$ by a transformation in which the velocity approaches that of light; in other words, the corresponding three-momentum tends to infinity. This suggests that the relation between the two quantization schemes is

that between the usual Feynman–Dyson rules and the '$p \to \infty$' rules (Weinberg 1966), i.e. they are equivalent.

3. The retarded commutators of two local operators $A$, $B$ are respectively ordered along the time direction and along the normal to the light plane in the two schemes:

$$R_0(x) = [A(x), B(0)]\theta(x^0),$$

$$R_+(x) = [A(x), B(0)]\theta(x^+).$$

However, for $x^2 \geqslant 0$, a condition ensured by micro-causality, the two are the same.

We now consider a quark–gluon field theory on the light-plane (Jackiw 1972a). Using the notations $q(x)$ and $B(x)$ for the fields of the quarks and of the gluon respectively, we may write our Lagrangian density as

$$\mathscr{L} = \bar{q}\left\{\left(\frac{i}{2}\overleftrightarrow{\partial}_\mu - gB_\mu\right)\gamma^\mu - \mathscr{M}\right\}q - \tfrac{1}{4}B_{\mu\nu}B^{\mu\nu}. \tag{5.80}$$

In eqn (5.80) the gluon has been chosen to be massless and $\mathscr{M}$ is the quark mass matrix in the internal symmetry space. It is convenient to define two projection operators $P_+ = \frac{1}{2}\gamma^-\gamma^+$, $P_- = \frac{1}{2}\gamma^+\gamma^-$ which project the quark field $q(x)$ into $q_+(x)$ and $q_-(x)$ respectively, i.e. $q_\pm(x) = P_\pm q(x)$. Because of certain properties of the projection operators, namely,

$$P_\pm = \tfrac{1}{2}(1 \pm \gamma^0\gamma^3), \qquad P_+P_- = 0 = P_-P_+, \qquad P_+ + P_- = 1,$$

we can take one of the projected fields, say $q_+$, as independent; $q_-$ then is a dependent field. The equation of motion may be handled in the following way to yield a constraint equation on the latter. Rewrite

$$(i\gamma\,.\,\partial - g\not{B})q = \mathscr{M}q$$

as

$$(i\gamma^+\partial_+ + i\gamma^-\partial_- + i\gamma_r\partial^r - g\gamma^+B^- - g\gamma^-B^+ - g\gamma_rB^r)q = \mathscr{M}q.$$

The left-multiplication of this equation by $\frac{1}{2}\gamma^+$ and the subsequent use of the relations $(\gamma^+)^2 = 0$, $\gamma^+q = \gamma^+q_+$ lead to the constraint equation

$$(i\partial_- - gB^+)q_- = \tfrac{1}{2}\{(i\partial_r - gB_r)\gamma^r + \mathscr{M}\}\gamma^+q_+. \tag{5.81}$$

Because the present gluon field is massless (or, even if it were massive, by virtue of its coupling to a conserved current), there exists a gauge freedom, with the aid of which $B^+$ may be set to zero. Eqn (5.81) may then be integrated readily; in the chosen gauge the dependent field is

$$q_-(x) = -\frac{i}{4}\int_{-\infty}^{\infty} d\xi\, \varepsilon(x^- - \xi)[\{i\partial_r - gB_r(x^+, \mathbf{x}_\perp, \xi)\}\gamma^r + \mathscr{M}] \times$$

$$\times \gamma^+ q_+(x^+, \mathbf{x}_\perp, \xi). \tag{5.82}$$

We now have to specify the canonical commutators. That between two gluon fields may be obtained in analogy with the corresponding equal-time commutators as

$$[\partial_- B^r(x), B^s(0)]_{x^+=0} = \frac{i}{2} g^{rs}\delta(x^-)\delta^{(2)}(\mathbf{x}_\perp), \tag{5.83a}$$

or

$$[B^r(x), B^s(0)]_{x^+=0} = \frac{i}{4} g^{rs}\varepsilon(x^-)\delta^{(2)}(\mathbf{x}_\perp). \tag{5.83b}$$

Eqn (5.83b) is the integrated form of eqn (5.83a). The correctness of the factor of $\frac{1}{4}$ on the right-hand side of the former may be checked by taking the vacuum expectation value (see Exercise 5.3) of the equation. The canonical commutation relation between two independent quark fields may be obtained (see Exercise 5.4) by starting from the relation

$$\{\bar{q}(x), q(0)\} = (i\gamma\,.\,\partial + \mathscr{M})\,\Delta(x),$$

and is

$$\{q_+(x), q_+^\dagger(0)\}_{x^+=0} = \frac{1}{\sqrt{2}} P_+\delta(x^-)\delta^{(2)}(\mathbf{x}_\perp). \tag{5.84}$$

Apart from eqns (5.83) and (5.84), we have the trivial relations

$$0 = \{q_+(x), q_+(0)\}_{x^+=0} = \{q_+^\dagger(x), q_+^\dagger(0)\}_{x^+=0}$$
$$= [q_+(x), B^r(0)]_{x^+=0} = [q_+^\dagger(x), B^r(0)]_{x^+=0}.$$

Moreover, eqns (5.84) and (5.82) lead immediately to the result

$$\{q_-(x), q_+^\dagger(0)\}_{x^+=0} = -\frac{i}{4\sqrt{2}}\varepsilon(x^-)[\{i\partial_r - gB_r(0)\}\gamma^r + \mathscr{M}]\gamma^+\delta^{(2)}(\mathbf{x}_\perp). \tag{5.85}$$

Eqn (5.85) may also be verified in the manner of eqn (5.83b)

We now wish to evaluate current commutators on the light-plane for the theory introduced in the preceding paragraph. The components of the standard SU(3) × SU(3) vector and axial vector currents may be written as

$$
\begin{aligned}
J^{i,+} &= \sqrt{(2)}q_+^\dagger \frac{\lambda^i}{2} q_+ ,\\
J^{i,-} &= \sqrt{(2)}q_-^\dagger \frac{\lambda^i}{2} q_- ,\\
J^{i,r} &= \frac{1}{\sqrt{2}} q_+^\dagger \frac{\lambda^i}{2}\gamma^-\gamma^r q_- + \frac{1}{\sqrt{2}} q_-^\dagger \frac{\lambda^i}{2}\gamma^+\gamma^r q_+ ,\\
J_5^{i,+} &= \sqrt{(2)}q_+^\dagger \frac{\lambda^i}{2}\gamma_5 q_+ ,\\
J_5^{i,-} &= \sqrt{(2)}q_-^\dagger \frac{\lambda^i}{2}\gamma_5 q_- ,\\
J_5^{i,r} &= \frac{1}{\sqrt{2}} q_+^\dagger \frac{\lambda^i}{2}\gamma^-\gamma^r\gamma_5 q_- + \frac{1}{\sqrt{2}} q_-^\dagger \frac{\lambda^i}{2}\gamma^+\gamma^r\gamma_5 q_+ .
\end{aligned}
\tag{5.86}
$$

Armed with these quark representations of the currents and with the basic anticommutators (eqns (5.84) and (5.85)), we may now obtain the commutation relations among the currents much in the manner of eqns (5.36). The final results for the vector currents are (see Exercise (5.5))

$$
[J^{i,+}(x), J^{j,+}(0)]_{x^+=0} = \mathrm{i}f^{ijk}J^{k,+}(x)\delta(x^-)\delta^{(2)}(\mathbf{x}_\perp) + \text{D.P.}, \tag{5.87a}
$$

$$
\begin{aligned}
[J^{i,+}(x), J^{j,r}(0)]_{x^+=0} = {} & \mathrm{i}f^{ijk}J^{k,r}(x)\delta(x^-)\delta^{(2)}(\mathbf{x}_\perp) - \\
& -\tfrac{1}{8}(\mathrm{i}f^{ijk} + d^{ijk})\big(\partial_-[\varepsilon(x^-)\delta^{(2)}(\mathbf{x}_\perp)\{J^{k,r}(x,0) - \mathrm{i}\varepsilon^{rs}J_{s5}(x,0)\}] + \\
& + \partial_s[\varepsilon(x^-)\delta^{(2)}(\mathbf{x}_\perp)\{-g^{rs}J^{k,+}(x,0) + \mathrm{i}\varepsilon^{rs}J_5^{k,+}(x,0)\}]\big) + \\
& + \frac{\mathrm{i}}{4}\varepsilon(x^-)\delta^{(2)}(\mathbf{x}_\perp)\bar{O}^{ij,r}(x,0) - \text{h.c.} + \text{D.P.},
\end{aligned}
\tag{5.87b}
$$

$$
\begin{aligned}
[J^{i,+}(x), J^{j,-}(0)]_{x^+=0} = {} & \mathrm{i}f^{ijk}J^{k,-}(x)\delta(x^-)\delta^{(2)}(\mathbf{x}_\perp) - \\
& -\tfrac{1}{4}(\mathrm{i}f^{ijk} + \mathrm{d}^{ijk})\big(\partial_-\{\varepsilon(x^-)\delta^{(2)}(\mathbf{x}_\perp)J^{k,-}(x,0)\} + \\
& + \tfrac{1}{2}\partial_r[\varepsilon(x^-)\delta^{(2)}(\mathbf{x}_\perp)\{J^{k,r}(x,0) + \mathrm{i}\varepsilon^{rs}J_{s5}^k(x,0)\}]\big) + \\
& + \frac{\mathrm{i}}{4}\varepsilon(x^-)\delta^{(2)}(\mathbf{x}_\perp)O^{ij}(x,0) - \text{h.c.} + \text{D.P.}.
\end{aligned}
\tag{5.87c}
$$

In eqns (5.87) we have used the bilocal currents $J_\mu^k(x, 0)$, $J_{\mu 5}^k(x, 0)$ which were introduced previously. $O^{ij}(x, 0)$ and $\bar{O}^{ij,r}(x, 0)$ are two bilocal operators with the following quark structure:

$$O^{ij}(x, 0) \sim \bar{q}(x)\gamma^+\gamma^-\left[\mathcal{M}, \frac{\lambda^i}{2}\right]\frac{\lambda^j}{2}q(0), \tag{5.88a}$$

$$\bar{O}^{ij,r}(x, 0) \sim \bar{q}(x)\gamma^+\gamma^r\left[\mathcal{M}, \frac{\lambda^i}{2}\right]\frac{\lambda^j}{2}q(0). \tag{5.88b}$$

Moreover, we have displayed only the connected contributions to the commutators, collecting the disconnected parts under D.P. Results similar to eqns (5.87) may be obtained with axial vector currents (see Exercise 5.4).

*Comments on light-plane current commutators*

The following observations may be made on the light-plane commutation relations of eqns (5.87).

1. Of the three commutators in eqns (5.87), the $++$ one of eqn (5.87a) is determined solely by the fundamental quantization condition of eqn (5.84), without the use of the equation of motion (eqn (5.82)). The other two commutators depend on eqn (5.85) and hence on eqn (5.82), and are more model-dependent.

2. The greater model-dependence of eqns (5.87c) and (5.87b) is shown also by the presence of the operators $O^{ij}(x, 0)$ and $\bar{O}^{ij,r}(x, 0)$. These involve the commutator $[\mathcal{M}, \lambda^i]$ and hence depend on the details of the breaking of SU(3) symmetry.

3. The terms left unspecified under D.P. all involve total derivatives with respect to $x^-, x^r$ or both (Jackiw 1972a). Formal considerations imply that they are disconnected $c$-numbers. Henceforth we shall assume the absence of non-canonical effects which might make them operators.

4. In the gauge chosen ($B^+ = 0$), the bilocal currents have the free-field quark structure. However, their gauge-invariant forms are

$$\bar{q}(0, \mathbf{0}_\perp, x^-)\gamma^\mu q(0) \exp\left\{ig\int_0^{x^-} \mathrm{d}y^-\, B^+(0, \mathbf{0}_\perp, y^-)\right\}, \tag{5.89a}$$

$$\bar{q}(0, \mathbf{0}_\perp, x^-)\gamma^\mu\gamma_5 q(0) \exp\left\{ig\int_0^{x^-} \mathrm{d}y^-\, B^+(0, \mathbf{0}_\perp, y^-)\right\} \tag{5.89b}$$

for the vector and axial vector cases respectively. Eqns (5.89) are obtained from eqn (5.47) and from the corresponding axial relation by considering the edge of intersection between the light-cone and the light-plane.

5. The bilocal currents described in eqn (5.89) also satisfy a closed algebra on the light-plane. Of these algebraic relations those which involve $++$ components (known as 'good–good' commutators) are supposed to be

exact as a consequence of light-plane quantization. The others are less reliable and, at most, may be true to the leading order on the light-plane. For details the reader is referred to the most recent literature (Fritzsch and Gell-Mann 1972).

## Exercises

5.1. Show that $\nu W_2^e$ scales in the Bjorken limit by deriving eqn (5.30).

5.2. Show the consistency of light-cone current algebra with current conservation (eqn (5.41)) to leading order on the light-cone, without explicitly using the massless quark-field model.

5.3. Verify eqn (5.83b) by taking the vacuum expectation value of either side and writing a spectral representation for the commutator. Similarly verify eqn (5.85).

5.4. Derive the canonical anticommutator of eqn (5.84) by following the suggestion given in the text.

5.5. Prove the results

$$\gamma^r P_\pm = \tfrac{1}{2}\gamma^r \mp \frac{i}{2}\varepsilon^{rs}\gamma_s\gamma_5,$$

$$0 = \delta(x^-)\delta^{(2)}(\mathbf{x}_\perp)\{q^\dagger(x)Aq(0) - q^\dagger(0)Aq(x)\}_{x^+=0},$$

where $A$ is any combination of gamma matrices. Make use of these relations in deriving eqns (5.87) and the corresponding commutators involving axial currents.

# 6

# APPLICATIONS OF LIGHT-CONE PHYSICS

## Spin-averaged deep inelastic *l*N scattering

THE motivation for the development of light-cone current algebra was to incorporate Bjorken scaling within a theoretical framework of hadronic currents consistent with the already familiar equal-time current algebra. This the scheme achieves. As an added bonus all the 'reliable' results of the quark–parton model (Chapter 1) are also obtained.

*Scale functions*

We start with the absorptive part of the forward amplitude for the scattering of a hadronic vector current from a spin-averaged nucleon target:

$$W^{ij}_{\mu\nu} \equiv (2\pi)^2 \frac{p_0}{M} \int \mathrm{d}^4 z\, \mathrm{e}^{\mathrm{i}q.z} \left\langle \mathrm{N} \left| \left[ J^i_\mu\left(\frac{z}{2}\right), J^j_\nu\left(-\frac{z}{2}\right)\right] \right| \mathrm{N} \right\rangle$$

$$= \left(-g_{\mu\nu} + \frac{q_\mu q_\nu}{q^2}\right) W^{ij}_1(q^2, \nu) + \frac{1}{M^2}\left(p_\mu - \frac{p\cdot q}{q^2} q_\mu\right)\left(p_\nu - \frac{p\cdot q}{q^2} q_\nu\right) W^{ij}_2(q^2, \nu). \tag{6.1}$$

When we go to the light-cone and use eqn (5.36) in the above equation, we obtain both vector and axial bilocal currents from the commutator. The axial currents do not contribute to the forward spin-averaged matrix elements of single-particle states. Thus we have

$$W^{ij}_{\mu\nu} \triangleq 2\pi \frac{p_0}{M} \int \mathrm{d}^4 z\, \mathrm{e}^{\mathrm{i}q.z} \partial^\rho_z \{\varepsilon(z^0)\delta(z^2)\} S_{\mu\rho\nu\sigma} \times$$
$$\times \left\langle \mathrm{N} \left| \mathrm{i} f^{ijk} J^{\sigma,\oplus,k}\left(\frac{z}{2}, -\frac{z}{2}\right) + \mathrm{d}^{ijk} J^{\sigma,\ominus,k}\left(\frac{z}{2}, -\frac{z}{2}\right) \right| \mathrm{N} \right\rangle. \tag{6.2}$$

Since the light-cone commutator of two hadronic vector currents is the same as that of the corresponding axial vector currents, we note that the right-hand side of eqn (6.2) also describes the leading light-cone limit of the structure tensor for which the vector currents $J^i_\mu$ and $J^j_\nu$ of eqn (6.1) are replaced by $J^i_{\mu 5}$ and $J^j_{\nu 5}$ respectively. Consider now the quark structure of the matrix element of the symmetric bilocal current in eqn (6.2),

$$\left\langle \mathrm{N} \left| J^{\sigma,\oplus,k}\left(\frac{z}{2}, -\frac{z}{2}\right) \right| \mathrm{N} \right\rangle$$

$$\sim \frac{1}{2} \left\langle \mathrm{N} \left| \bar{q}\left(\frac{z}{2}\right) \gamma^\sigma \frac{\lambda^k}{2} q\left(-\frac{z}{2}\right) + \bar{q}\left(-\frac{z}{2}\right) \gamma^\sigma \frac{\lambda^k}{2} q\left(\frac{z}{2}\right) \right| \mathrm{N} \right\rangle$$

$$= \frac{1}{2}\left\{\left\langle N \left| 2\bar{q}(0)\gamma^{\sigma}\frac{\lambda^k}{2}q(0) + \frac{2z^{\alpha}z^{\beta}}{2!4}\left(\bar{q}\gamma^{\sigma}\frac{\lambda^k}{2}\partial_{\alpha}\partial_{\beta}q + \partial_{\alpha}\partial_{\beta}\bar{q}\gamma^{\sigma}\frac{\lambda^k}{2}q - \right.\right.\right.\right.$$
$$\left.\left.\left.\left. -2\partial_{\alpha}\bar{q}\gamma^{\sigma}\frac{\lambda^k}{2}\partial_{\beta}q\right) + \cdots \right| N\right\rangle\right\}.$$

Since the symmetric bilocal current is even under the replacement $z \leftrightarrow -z$, only even powers of $z$ contribute. Similarly, only odd powers of $z$ contribute to the matrix element of the antisymmetric bilocal current and we have

$$\left\langle N \left| J^{\sigma,\ominus,k}\left(\frac{z}{2}, -\frac{z}{2}\right)\right| N\right\rangle$$
$$\sim \frac{1}{2}\left\langle N \left| z^{\alpha}\left(-\bar{q}\gamma^{\sigma}\frac{\lambda^k}{2}\partial_{\alpha}q + \partial_{\alpha}\bar{q}\gamma^{\sigma}\frac{\lambda^k}{2}q\right) + \right.\right.$$
$$+\frac{2z^{\alpha}z^{\beta}z^{\gamma}}{3!8}\left(-\bar{q}\gamma^{\sigma}\frac{\lambda^k}{2}\partial_{\alpha}\partial_{\beta}\partial_{\gamma}q + 3\partial_{\alpha}\bar{q}\gamma^{\sigma}\frac{\lambda^k}{2}\partial_{\beta}\partial_{\gamma}q - 3\partial_{\alpha}\partial_{\beta}\bar{q}\gamma^{\sigma}\frac{\lambda^k}{2}\partial_{\gamma}q + \right.$$
$$\left.\left. +\partial_{\alpha}\partial_{\beta}\partial_{\gamma}\bar{q}\gamma^{\sigma}\frac{\lambda^k}{2}q\right) + \cdots \right| N\right\rangle.$$

We can parametrize these matrix elements more generally as

$$\left\langle N \left| J_{\sigma}^{\oplus,k}\left(\frac{z}{2}, -\frac{z}{2}\right)\right| N\right\rangle \triangleq \left(\frac{1}{2\pi}\right)^3 \frac{1}{2p_0}\left\{\tilde{S}_0^k p_{\sigma} + \frac{\tilde{S}_2^k}{2!}(z\,.\,p)^2 p_{\sigma} + \cdots\right\}$$
$$= \left(\frac{1}{2\pi}\right)^3 \frac{1}{2p_0}\tilde{S}^k(p\,.\,z)p_{\sigma}, \qquad (6.3)$$

$$\left\langle N \left| J_{\sigma}^{\ominus,k}\left(\frac{z}{2}, -\frac{z}{2}\right)\right| N\right\rangle \triangleq \left(\frac{1}{2\pi}\right)^3 \frac{1}{2p_0}\left\{\tilde{A}_1^k p_{\sigma}(z\,.\,p) + \frac{\tilde{A}_3^k}{3!}z\,.\,p)^3 p_{\sigma} + \cdots\right\}$$
$$= \left(\frac{1}{2\pi}\right)^3 \frac{1}{2p_0}\tilde{A}^k(z\,.\,p)p_{\sigma}, \qquad (6.4)$$

where $\tilde{S}^k(p\,.\,z)$ and $\tilde{A}^k(p\,.\,z)$ are even and odd functions of $p\,.\,z$ respectively. In obtaining eqns (6.3) and (6.4), we have taken the leading terms of the matrix elements of the $z$-independent tensor operators to be proportional to $p_{\sigma}$, $p_{\sigma}p_{\alpha}$, $p_{\sigma}p_{\alpha}p_{\beta}$, etc., because any other term, such as $M^2 g_{\sigma\alpha}$ (called a 'trace term'), must contain the metric tensor at least once and can only generate non-leading contributions. This will become clearer as we go on. We take the leading matrix elements, as given by eqns (6.3) and (6.4), and substitute in eqn (6.2). This changes the structure tensor to

$$W_{\mu\nu}^{ij} \triangleq \frac{S_{\mu}{}^{\rho}{}_{\nu}{}^{\sigma}p_{\sigma}}{8\pi^2 M}\int d^4z\, e^{iq.z}\partial_{\rho}\{\varepsilon(z^0)\delta(z^2)\}\{if^{ijk}\tilde{S}^k(z\,.\,p) + d^{ijk}\tilde{A}^k(z\,.\,p)\}. \qquad (6.5)$$

A similar expression can be obtained for the corresponding 'axial' structure tensor $W_{\mu\nu 5}$ which generates the parity-violating terms in inelastic neutrino and antineutrino scattering, namely,

$$W^{ij}_{\mu\nu 5} \equiv (2\pi)^2 \frac{p_0}{M} \int \mathrm{d}^4 z\, \mathrm{e}^{\mathrm{i}q.z} \langle \mathrm{N} | [J^i_{\mu 5}(z), J^j_\nu(0)] | \mathrm{N} \rangle$$

$$\triangleq \frac{\mathrm{i}\varepsilon_{\mu\nu\rho\sigma} p^\sigma}{8\pi^2 M} \int \mathrm{d}^4 z\, \mathrm{e}^{\mathrm{i}q.z} \partial^\rho \{\varepsilon(z^0)\delta(z^2)\} \times$$

$$\times \{\mathrm{i} f^{ijk} \tilde{A}^k(p\,.\,z) - \mathrm{d}^{ijk} \tilde{S}^k(p\,.\,z)\}. \tag{6.6}$$

We now introduce the Fourier transforms $S^k(\alpha)$, $A^k(\alpha)$ of the even and odd functions $\tilde{S}^k(z\,.\,p)$ and $\tilde{A}^k(z\,.\,p)$ respectively,

$$\tilde{S}^k(z\,.\,p) = \int_{-\infty}^{\infty} \mathrm{d}\alpha\, \mathrm{e}^{\mathrm{i}\alpha z.p} S^k(\alpha), \tag{6.7a}$$

$$\tilde{A}^k(z\,.\,p) = \int_{-\infty}^{\infty} \mathrm{d}\alpha\, \mathrm{e}^{\mathrm{i}\alpha z.p} A^k(\alpha). \tag{6.7b}$$

Thus eqn (6.5) may be rewritten as

$$W^{ij}_{\mu\nu} \triangleq \frac{S_\mu{}^\rho{}_\nu{}^\sigma p_\sigma}{8\pi^2 M} \int_{-\infty}^{\infty} \mathrm{d}\alpha \int \mathrm{d}^4 z\, \mathrm{e}^{\mathrm{i}(q+\alpha p).z} \partial_\rho \{\varepsilon(z^0)\delta(z^2)\} \times$$

$$\times \{\mathrm{i} f^{ijk} S^k(\alpha) + d^{ijk} A^k(\alpha)\}$$

$$= \frac{-S_\mu{}^\rho{}_\nu{}^\sigma p_\sigma}{8\pi^2 M} \int_{-\infty}^{\infty} \mathrm{d}\alpha (q+\alpha p)_\rho \int \mathrm{d}^4 z . \varepsilon^{\mathrm{i}(q+\alpha p).z} \varepsilon(z^0)\delta(z^2) \times$$

$$\times \{\mathrm{i} f^{ijk} S^k(\alpha) + \mathrm{d}^{ijk} A^k(\alpha)\}, \tag{6.8}$$

where we have discarded a surface term.

Using the relation

$$\int \mathrm{d}^4 z\, \varepsilon(z^0)\delta(z^2)\, \mathrm{e}^{\mathrm{i}(q+\alpha p).z} = -\frac{4\pi^2}{\mathrm{i}} \delta\{(q+\alpha p)^2\} \varepsilon(q^0 + p^0) \tag{6.9}$$

in eqn (6.8), we obtain

$$W^{ij}_{\mu\nu} \triangleq S_\mu{}^\rho{}_\nu{}^\sigma p_\sigma \frac{(q+wp)_\rho}{4q\,.\,pM} \{\mathrm{i} f^{ijk} S^k(w) + d^{ijk} A^k(w)\}. \tag{6.10}$$

The corresponding equation for the 'axial' structure tensor is

$$W^{ij}_{\mu\nu 5} \triangleq \mathrm{i}\varepsilon_{\mu\nu\rho\sigma} p^\sigma \frac{(q+wp)^\rho}{4q\,.\,pM} \{\mathrm{i} f^{ijk} A^k(w) - d^{ijk} S^k(w)\}. \tag{6.11}$$

We can now see why the 'trace terms' in the matrix elements of the $z$-independent tensors, to which we alluded earlier, are really non-leading in the limit concerned. Take

$$\left\langle \mathrm{N} \middle| J_\sigma^{\ominus,k}\left(\frac{z}{2}, -\frac{z}{2}\right) \middle| \mathrm{N} \right\rangle (2\pi)^3 2p_0 = \tilde{A}^k(p\,.\,z)p_\sigma + \bar{A}^k(p\,.\,z)z_\sigma,$$

and similarly for the symmetric bilocal current. The contribution of the second term in the right-hand side above to the quantity of interest can be evaluated to be proportional to

$$MS_\mu{}^\rho{}_\nu{}^\sigma \frac{\partial}{\partial q^\sigma} \int_{-\infty}^{\infty} \mathrm{d}\alpha\,(q+\alpha p)_\rho \int \mathrm{d}^4 z\, \mathrm{e}^{\mathrm{i}(q+\alpha p).z} \varepsilon(z_0)\delta(z^2) \times$$

$$\times \{\mathrm{i}f^{ijk}\bar{S}^k(\alpha) + d^{ijk}\bar{A}^k(\alpha)\}$$

$$\propto MS_\mu{}^\rho{}_{\nu\rho} \frac{1}{p\,.\,q}\{\mathrm{i}f^{ijk}\bar{S}^k(w) + d^{ijk}\bar{A}^k(w)\} +$$

$$+\text{similar or less-leading terms}$$

$$= -\frac{2}{\nu}\{\mathrm{i}f^{ijk}\bar{S}^k(w) + d^{ijk}\bar{A}^k(w)\}g_{\mu\nu} +$$

$$+\text{similar or less-leading terms.}$$

Thus the extra term gives contributions which go to zero in the Bjorken limit as $\nu^{-1}$ times a function of $w$. Returning to eqns (6.10) and (6.11), we evaluate the scale functions. Comparing the coefficients of the $p_\mu p_\nu M^{-2}$ terms in eqns (6.1) and (6.10), we are led to

$$\lim_{\mathrm{Bj}} \nu W_2^{ij} = F_2^{ij}(w) = \frac{w}{2}\{\mathrm{i}f^{ijk}S^k(w) + d^{ijk}A^k(w)\}. \tag{6.12}$$

A similar comparison of the $g^{\mu\nu}$ term leads to (see Exercise 6.1)

$$\lim_{\mathrm{Bj}} MW_1^{ij} = F_1^{ij}(w) = \frac{1}{2w}F_2^{ij}(w), \tag{6.13}$$

i.e. vanishing longitudinal cross-sections $\sigma_{\mathrm{L}}$ (see Introduction, p. 20) in the Bjorken limit. In fact it follows from the above considerations that (see Exercise 6.2) the ratio $R \equiv \sigma_{\mathrm{L}}/\sigma_{\mathrm{T}}$ tends to zero in the Bjorken limit as $\nu^{-1}$ times a function of $w$. Finally, we can make a similar comparison for the $\varepsilon_{\mu\nu\rho\sigma}q^\rho p^\sigma$ term of $W_{\mu\nu 5}^{ij}$ in obtaining the Bjorken limit of the 'parity-violating' structure function as (see Exercise 6.3)

$$\lim_{\mathrm{Bj}} \nu W_3^{ij} = F_3^{ij}(w) = -\tfrac{1}{2}\{\mathrm{i}f^{ijk}A^k(w) + d^{ijk}S^k(w)\}. \tag{6.14}$$

*Internal symmetry relations*

We shall now discuss the consequences of internal symmetries. In the electromagnetic case the SU(3) current index $i = (3)+1/\sqrt{3}(8)$. In the weak case we may take $i = (1)\pm i(2)$ provided we approximate the Cabibbo angle $\theta_C$ to zero and change the vector current to the V–A form, i.e. $J_\mu \to J_\mu - J_{\mu 5}$. Working out the SU(3) Clebsch–Gordan coefficients (Gell-Mann and Nee'man (1964)) and including the extra factor of 2 for the weak case, we obtain

$$F_2^{\mathrm{ep}}(w) = w\left\{\tfrac{2}{3}\sqrt{(\tfrac{2}{3})}A_\mathrm{p}^0(w)+\tfrac{1}{3}A_\mathrm{p}(w)+\frac{1}{3\sqrt{3}}A_\mathrm{p}^8(w)\right\}, \tag{6.15}$$

$$F_2^{\substack{\nu\mathrm{p}\\ \bar{\nu}\mathrm{p}}}(w) = w\left\{2\sqrt{(\tfrac{2}{3})}A_\mathrm{p}^0(w)\mp 2S_\mathrm{p}^3(w)+\frac{2}{\sqrt{3}}A_\mathrm{p}^8(w)\right\}, \tag{6.16}$$

$$F_3^{\substack{\nu\mathrm{p}\\ \bar{\nu}\mathrm{p}}}(w) = -2\sqrt{(\tfrac{2}{3})}S_\mathrm{p}^0(w)\pm 2A_\mathrm{p}^3(w)-\frac{2}{\sqrt{3}}S_\mathrm{p}^8(w). \tag{6.17}$$

The superscript or subscript p in the right-hand sides of eqns (6.15)–(6.17) refers to the target proton. We note that isospin symmetry implies

$$A_\mathrm{p}^{0,8} = A_\mathrm{n}^{0,8},\ A_\mathrm{p}^3 = -A_\mathrm{n}^3. \tag{6.18}$$

Eqns (6.15)–(6.18) lead to the Llewellyn-Smith equality

$$6\{F_2^{\mathrm{ep}}(w)-F_2^{\mathrm{en}}(w)\} = 4A_\mathrm{p}^3(w) = w\{F_3^{\nu\mathrm{p}}(w)-F_3^{\bar{\nu}\mathrm{p}}(w)\}, \tag{6.19}$$

obtained previously in the quark–parton model (Chapter 1). In addition, we have the integral relation

$$\int_{-1}^{1}\frac{\mathrm{d}w}{w}\{F_2^{\bar{\nu}\mathrm{p}}(w)-F_2^{\nu\mathrm{p}}(w)\} = 4\int_{-1}^{1}\mathrm{d}w\ S_\mathrm{p}^3(w) = 4\tilde{S}_\mathrm{p}^3(0) = 4. \tag{6.20}$$

The last step in eqn (6.20), i.e. $\tilde{S}_\mathrm{p}^3(0) = 1$, follows from the relation

$$p_\sigma \tilde{S}_\mathrm{p}^3(0) = (2\pi)^3 2p_0\langle p|J_\sigma^3(0)|\mathrm{p}\rangle = p_\sigma$$

by the conserved vector current (CVC) hypothesis (Marshak, Riazuddin, and Ryan 1969). Using the evenness of $F_2^{\bar{\nu}\mathrm{p}}(w)-F_2^{\nu\mathrm{p}}(w)$ in eqn (6.20), we obtain the Adler sum rule (CCSR)

$$\int_0^1\frac{\mathrm{d}w}{w}\{F_2^{\bar{\nu}\mathrm{p}}(w)-F_2^{\nu\mathrm{p}}(w)\} = 2. \tag{6.21a}$$

Similarly, we obtain

$$\int_{-1}^{1} \mathrm{d}w\{F_3^{\bar{\nu}p}(w)+F_3^{\nu p}(w)\} = -4\int_{-1}^{1} \mathrm{d}w\left\{\sqrt{(\tfrac{2}{3})}S_p^0(w)+\frac{1}{\sqrt{3}}S_p^8(w)\right\}$$

$$= -4\left\{\sqrt{(\tfrac{2}{3})}\tilde{S}_p^0(0)+\frac{1}{\sqrt{3}}\tilde{S}_p^8(0)\right\} = -8\left(B_p+\frac{Y_p}{2}\right) = -12$$

or

$$\int_0^1 \frac{\mathrm{d}w}{w}\{F_3^{\nu p}(w)+F_3^{\bar{\nu}p}(w)\} = -6. \tag{6.21b}$$

Eqn (6.21b) is the Gross–Llewellyn-Smith sum rule (BCSR), as discussed in the quark–parton model (Chapter 1).

An interesting sum rule obtains in the absence of gluons. If there are no gluons, we can write the quark structure of the stress–energy tensor as

$$\theta_{\mu\nu} \sim \frac{1}{4\mathrm{i}}(\bar{q}\gamma_\mu\partial_\nu q-\partial_\nu\bar{q}\gamma_\mu q+\mu\leftrightarrow\nu). \tag{6.22}$$

From eqns (6.15)–(6.18), we may write

$$6\{F_2^{ep}(w)+F_2^{en}(w)\}-\{F_2^{\nu p}(w)+F_2^{\bar{\nu}p}(w)\} = 4\sqrt{(\tfrac{2}{3})}A_p^0(w)w. \tag{6.23}$$

However, it follows from the definition of $A^0(w)$ and eqns (6.4) and (6.7) that

$$A^0(w) = \frac{1}{2\pi}\int_{-\infty}^{\infty} \mathrm{d}(z\,.\,p)\,\mathrm{e}^{-\mathrm{i}wz.p}\tilde{A}^0(z\,.\,p)$$

$$= \frac{1}{2\pi}\int_{-\infty}^{\infty} \mathrm{d}(z\,.\,p)\,\mathrm{e}^{-\mathrm{i}wz.p}\left\{\tilde{A}_1^0 z\,.\,p+\frac{\tilde{A}_3^0}{3!}(z\,.\,p)^3+\ldots\right\}$$

$$= \mathrm{i}\left\{\tilde{A}_1^0\delta'(w)-\frac{A_3^0}{3!}\delta'''(w)+\ldots\right\}. \tag{6.24}$$

In eqn (6.24)

$$\left(\frac{1}{2\pi}\right)^3\frac{1}{2p_0}\tilde{A}_1^0 p_\sigma p_\alpha \sim \tfrac{1}{2}\langle \mathrm{N}|\bar{q}\gamma_\sigma\partial_\alpha q-\partial_\alpha\bar{q}\gamma_\sigma q|\mathrm{N}\rangle\tfrac{1}{2}\sqrt{(\tfrac{2}{3})},$$

etc. Thus, by using eqn (6.22), in the absence of gluons, we may take

$$\left(\frac{1}{2\pi}\right)^3\frac{1}{2p_0}\tilde{A}_1^0 p_\sigma p_\alpha \underset{\text{no gluons}}{=} \mathrm{i}\langle \mathrm{N}|\theta_{\sigma\alpha}|\mathrm{N}\rangle\tfrac{1}{2}\sqrt{(\tfrac{2}{3})}. \tag{6.25}$$

However, the stress–energy tensor has a universal matrix element between spin-averaged single-particle states:

$$(2\pi)^3 \frac{p_0}{M}\langle \mathrm{N}|\theta_{\mu\nu}|\mathrm{N}\rangle = \frac{p_\mu p_\nu}{M}. \qquad (6.26)$$

From eqns (6.25) and (6.26), we have

$$\tilde{A}_2^0 = \mathrm{i}\sqrt{(\tfrac{2}{3})},$$

which leads, by use of eqns (6.23) and (6.24), to the sum rule

$$\int_{-1}^{1} \mathrm{d}w\,[6\{F_2^{\mathrm{ep}}(w)+F_2^{\mathrm{en}}(w)\}-\{F_2^{\nu\mathrm{p}}(w)+F_2^{\bar{\nu}\mathrm{p}}(w)\}] = -\int_{-1}^{1} \mathrm{d}w\,\tfrac{8}{3}w\delta'(w) = \tfrac{8}{3},$$

or

$$\int_{0}^{1} \mathrm{d}w[6\{F_2^{\mathrm{ep}}(w)+F_2^{\mathrm{en}}(w)\}-\{F_2^{\nu\mathrm{p}}(w)+F_2^{\bar{\nu}\mathrm{p}}(w)\}] = \tfrac{4}{3}.$$

The above result should be compared with eqn (1.49)

$$\varepsilon_{\mathrm{g}} = 1-\tfrac{3}{4}\int_{0}^{1} \mathrm{d}w\{6F_2^{\mathrm{ep}+\mathrm{en}}(w)-F_2^{\nu\mathrm{p}+\bar{\nu}\mathrm{p}}(w)\},$$

where $\varepsilon_{\mathrm{g}}$ is an appropriately defined measure of the gluon content in a nucleon, derived from the parton model (Chapter 1). Other sum rules of the quark–parton model, using symmetric momentum distribution, etc. (see Table 1.1, p. 38), can be obtained here only by introducing assumptions about the matrix elements of the expansion operators on the light-cone. These certainly need postulates beyond the light-cone algebra of currents. Finally, we mention that the imposition of positivity constraints, coupled with isospin invariance, on the functions $S^k(w)$ and $A^k(w)$ leads to all the inequalities derived for the quark–parton model of Chapter 1, namely, the Llewellyn-Smith inequality, the Majumdar–Nachtmann inequality, and Nachtmann's neutrino inequalities. The method for deriving them is not very different from that used in the parton case, and will not be given here. The interested reader may find the derivations in the literature (Callan 1972b). It is, however, not without significance that the two approaches, quark–parton model and light-cone current algebra, lead to the same results in spin-averaged deep inelastic $l$N scattering.

### Spin-dependent deep inelastic scattering

In our discussions of $l$N scattering so far, we have always averaged over the spins of the target nucleon. We shall now consider the case of polarized targets. For definiteness, let us confine ourselves to electron scattering. Our considerations may be readily generalized to include neutrinos and antineutrinos.

*Kinematics*

We give only a short and simple presentation of the kinematics involved (Doncel and de Rafael 1971). For polarized target nucleons the usual structure tensor changes to

$$(2\pi)^2 \frac{p_0}{M} \int d^4x\, e^{iq.x} \langle N(p), s|[J_\mu^{EM}(x), J_\nu^{EM}(0)]|N(p), s\rangle = W_{\mu\nu}^e + S_{\mu\nu}^e, \tag{6.27}$$

where the new spin-dependent part $S_{\mu\nu}^e$ satisfies the constraints

$$\sum_s S_{\mu\nu}^e = 0, \qquad q^\mu S_{\mu\nu}^e = q^\nu S_{\mu\nu}^e = 0. \tag{6.28}$$

In eqn (6.27) use the spin projection operator $\frac{1}{2}(1+\gamma_5 \not{s})$, where $s$ is the spin pseudovector:

$$s^\mu = \frac{1}{M}[\mathbf{p}\,.\,\hat{\mathbf{n}},\, M\hat{\mathbf{n}} + (\mathbf{p}\,.\,\hat{\mathbf{n}})/(p_0+M)\mathbf{p}] = \frac{i}{M}\bar{u}(p)\gamma^\mu\gamma_5 u(p);$$

this goes to $(0, \hat{\mathbf{n}})$ in the rest frame, $\hat{\mathbf{n}}$ being a unit vector, and satisfies $s^2 = -1$ and $s.p = 0$. This leads to the following general form for $S_{\mu\nu}^e$:

$$S_{\mu\nu}^e = \mathrm{Tr}(\not{p}+M)\tfrac{1}{2}(1+\gamma_5\not{s})\omega_{\mu\nu}^e, \tag{6.29}$$

$\omega_{\mu\nu}^e$ being a $4\times 4$ matrix independent of $s$. Thus $S_{\mu\nu}^e$ must be linear in $s$. Further, according to eqn (6.27), it should satisfy the so-called 'hermiticity' property $S_{\mu\nu} = S_{\nu\mu}^*$. Consistent with time-reversal invariance and eqns (6.28), we can choose four different parity-conserving tensors for $S_{\mu\nu}^e$, namely,

(1) $i\varepsilon_{\mu\nu\alpha\beta}q^\alpha s^\beta$,

(2) $i\varepsilon_{\mu\nu\alpha\beta}q^\alpha p^\beta (q\,.\,s)$,

(3) $i\varepsilon_{\mu\nu\alpha\beta}p^\alpha s^\beta - \dfrac{i}{q\cdot p}(p_\nu\varepsilon_{\mu\rho\alpha\sigma} - p_\mu\varepsilon_{\nu\rho\alpha\sigma})q^\rho p^\alpha s^\sigma$,

(4) $i\varepsilon_{\mu\nu\alpha\beta}p^\alpha s^\beta - \dfrac{i}{q^2}(q_\nu\varepsilon_{\mu\rho\alpha\sigma} - q_\mu\varepsilon_{\nu\rho\alpha\sigma})q^\rho p^\alpha s^\sigma$. (6.30)

Note that a tensor proportional to

$$\frac{1}{p\cdot q}(p_\mu\varepsilon_{\nu\rho\alpha\beta} + p_\nu\varepsilon_{\mu\rho\alpha\beta}) - \frac{1}{q^2}(q_\mu\varepsilon_{\nu\rho\alpha\beta} + q_\nu\varepsilon_{\mu\rho\alpha\beta})s^\rho p^\alpha p^\beta,$$

although consistent with eqns (6.28), cannot satisfy the requirements of 'hermiticity' and T-invariance simultaneously, and is therefore excluded.

However, the four tensors of (6.30) are not all independent, owing to the constraint $s \,.\, p = 0$. There are actually only two independent tensors, and the most general form of $S^{\text{e}}_{\mu\nu}$ may be written (see Exercise 6.3) as

$$S^{\text{e}}_{\mu\nu} = \mathrm{i}\varepsilon_{\mu\nu\alpha\beta}M^{-1}q^{\alpha}s^{\beta}X^{\text{e}}_{1}(Q^2, \nu)+\mathrm{i}\varepsilon_{\mu\nu\alpha\beta}M^{-3}q^{\alpha}p^{\beta}q\,.\,sX^{\text{e}}_{2}(Q^2, \nu), \quad (6.31)$$

where $X^{\text{e}}_{1,2}$ are real structure functions with the same dimension as $W^{\text{e}}_{1,2}$.

The existence of two and only two mutually independent spin-dependent tensors may also be seen by a helicity analysis of the amplitude of present interest. Such an analysis is complicated by the fact that a polarized target is not in a definite helicity state. Consider forward Compton scattering of virtual photons from a polarized target (Fig. 38). In general, we have 36

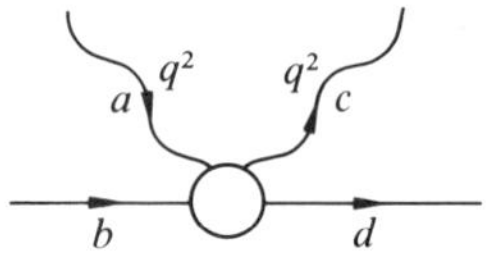


FIG. 38. Forward Compton scattering from a polarized target.

different helicity amplitudes. In the centre-of-mass system the situation is as illustrated in Fig. 39, where $a$, $b$, $c$, $d$ stand for appropriate helicities. Let $T_{ab,cd}$ be the helicity amplitude for the diagram. The conservation of angular momentum around the $z$-direction implies that $a-b = c-d$. Under the operation of time-reversal $T_{ab,cd} \rightarrow \mathscr{T}T_{ab,cd}\mathscr{T}^{-1} = T_{cd,ab}$ and under parity-inversion $T_{ab,cd} \rightarrow \mathscr{P}T_{ab,cd}\mathscr{P}^{-1} = T_{-a-b,-c-d}$. Therefore the invariance of the amplitude under these operations leaves us with only four amplitudes, which we take to be $T_{1\frac{1}{2},1\frac{1}{2}}$, $T_{0\frac{1}{2},0\frac{1}{2}}$, $T_{1-\frac{1}{2},1-\frac{1}{2}}$, and $T_{-1\frac{1}{2},0-\frac{1}{2}}$. These correspond to four structure functions, of which two are spin-dependent and two are not, thus giving us two tensors for $S^{\text{e}}_{\mu\nu}$.

We now come to the scaling properties of the new structure functions (Carlson and Tung 1972; Hey and Mandula 1972, Wray 1972). We shall not go into the kinematic formula for differential cross-sections off a polarized nucleon target (Doncel and de Rafael 1971), but we shall show one or two of the important results that do not need this formula. First, consider the polarization. If $\mathrm{d}\sigma_{\uparrow}$ is the differential cross-section with the nucleon polarized

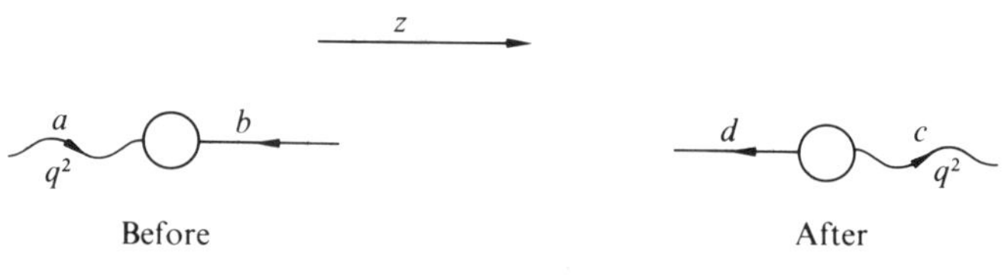


FIG. 39. Helicity in the centre-of-mass system for forward Compton scattering.

along the incident electron beam and $d\sigma_{\downarrow}$ stands for the differential cross-section with the opposite polarization, then the polarization $P$ is defined to be

$$P \equiv \frac{d\sigma_{\uparrow} - d\sigma_{\downarrow}}{d\sigma_{\uparrow} + d\sigma_{\downarrow}}. \tag{6.32}$$

Returning to eqn (I.1) and Fig. 2, we can construct a lepton tensor $L^{\mu\nu}$ as

$$L^{\mu\nu} = p_e^{\mu} p_e'^{\nu} + p_e'^{\mu} p_e^{\nu} - p_e \cdot p_e' g^{\mu\nu} + i\varepsilon^{\mu\nu\alpha\beta} p_{e\alpha} p'_{e\beta}. \tag{6.33}$$

The differential cross-section is to be obtained by contracting this with $W^e_{\mu\nu} + S^e_{\mu\nu}$. The contribution from the spin-dependent part is

$$L^{\mu\nu} S^e_{\mu\nu} = 2(-p_e \cdot q p'_e \cdot s + p_e \cdot s p'_e \cdot q) M^{-1} X^e_1 + 2(q \cdot s) \times$$
$$\times (-p_e \cdot q p'_e \cdot p + p'_e \cdot q p_e \cdot p) M^{-3} X^e_2. \tag{6.34}$$

In the laboratory frame (cf. the Introduction, p. 6)

$$p_e = (\varepsilon, 0, 0, \varepsilon), \qquad p'_e = (\varepsilon', 0, \varepsilon' \sin\theta, \varepsilon' \cos\theta),$$
$$q^0 = \varepsilon - \varepsilon' = \nu, \qquad Q^2 = 4\varepsilon\varepsilon' \sin^2 \tfrac{1}{2}\theta_e = -q^2, \qquad s = (0, \hat{\mathbf{n}}).$$

Choose the unit vector $\hat{\mathbf{n}}$ to be in the direction of the incident beam $\mathbf{p}_e$. Now, eqns (6.33) and (6.34) lead to

$$L^{\mu\nu} S^e_{\mu\nu} = -q^2(\varepsilon + \varepsilon' \cos\theta) M^{-1} X^e_1 - q^2 M^{-2} (\varepsilon + \varepsilon')(\varepsilon - \varepsilon' \cos\theta) X^e_2,$$

or

$$\lim_{\varepsilon \gg \nu} L^{\mu\nu} S^e_{\mu\nu} = -2q^2 \varepsilon M^{-1} (X^e_1 + \nu M^{-1} X^e_2).$$

Further, the spin-independent contribution from eqn (I.7) is

$$L^{\mu\nu} W^e_{\mu\nu} = q^2 W^e_1 + W^e_2 (2\varepsilon\varepsilon' + \tfrac{1}{2} q^2) \underset{\varepsilon \gg \nu}{\rightarrow} 2\varepsilon^2 W^e_2.$$

It follows from eqn (6.32) that, with the above choice of $n$, the polarization is

$$P = \frac{L^{\mu\nu} S^e_{\mu\nu}}{L^{\mu\nu} W^e_{\mu\nu}}.$$

Hence, in the limit when the incident energy $\varepsilon$ is much larger than the energy transfer $\nu$, we have

$$\lim P = \frac{1}{\varepsilon} \frac{X^e_1 + \nu X^e_2}{W_2} Q^2. \tag{6.35}$$

*Scaling and sum rules*

We now consider the application of light-cone current algebra to the present problem. The light-cone behaviour of the tensor $W^e_{\mu\nu} + S^e_{\mu\nu}$ may be written as

$$\lim_{\mathrm{LC}} (2\pi)^2 \frac{p_0}{M} \int \mathrm{d}^4 z\, \mathrm{e}^{\mathrm{i}q.z} \langle \mathrm{N}(p), s|[J_\mu^{i=(3)+(1/\surd 3)(8)}(z), J_\nu^{j=(3)+(1/\surd 3)(8)}(0)]|s, \mathrm{N}(p)\rangle$$

$$= (2\pi)^2 \frac{p_0}{M} \mathrm{d}^{ijk} \int \mathrm{d}^4 z\, \mathrm{e}^{\mathrm{i}q.z} \mathrm{i}\varepsilon_{\mu\nu\alpha\beta} \frac{1}{2\pi} \partial^\alpha \{\varepsilon(z^0)\delta(z^2)\} \times$$

$$\times \langle N(p), s|J_5^{\beta,\oplus,k}\left(\frac{z}{2}, \frac{z}{2}\right)|\mathrm{N}(p), s\rangle + W_{\mu\nu}^{\mathrm{e}} \text{ part.} \tag{6.36}$$

We have used eqn (5.36) in the last step of eqn (6.36) and retained only the d-coupling since $i = j = (3)+1/\surd 3(8)$ for the electromagnetic current (for the $(\nu, \bar{\nu})$-induced reactions the $f$-coupling will also contribute). Note that only the axial bilocal current is retained for the polarization-dependent part $S_{\mu\nu}^{\mathrm{e}}$ as opposed to the vector bilocal current which contributes to $W_{\mu\nu}^{\mathrm{e}}$, as discussed earlier. In our treatment we have been a little cavalier in using only the leading terms from light-cone current algebra and not retaining certain non-leading terms, as required by current conservation. Hence we shall have to systematically drop those terms which, to ensure current conservation, have to cancel with some of those which have been left out (Wray 1972). Now define the functions $\tilde{S}_{1,2}^k(p\,.\,z)$ by the relations

$$(2\pi)^3 \frac{p_0}{M} \left\langle \mathrm{N}(p), s\left| J_{\mu 5}^{\oplus,k}\left(\frac{z}{2}, -\frac{z}{2}\right)\right| s, \mathrm{N}(p)\right\rangle$$

$$= s_\mu \tilde{S}_1^k(p\,.\,z) + \mathrm{i}p_\mu s\,.\,z \tilde{S}_2^k(p\,.\,z) + \cdots. \tag{6.37}$$

In eqn (6.37) the dots stand for non-leading terms and we have used the requirement that the matrix element be linear in $s$. Since $J_\mu^\oplus(z/2, -z/2)$ is the symmetric bilocal current, the functions $\tilde{S}_1, \tilde{S}_2$ are even and odd in $z$ respectively. We now have to consider the following two terms in the Fourier integral in the right-hand side of eqn (6.36):

(1)
$$I_1^{\alpha\beta} = \int \mathrm{d}^4 z\, \mathrm{e}^{\mathrm{i}q.z} \partial^\alpha \{\varepsilon(z^0)\delta(z^2)\} \tilde{S}_1^k(p\,.\,z) s^\beta$$

$$= -\mathrm{i} \int_{-\infty}^{\infty} \mathrm{d}\xi\, (q+\xi p)^\alpha \int \mathrm{d}^4 z\, \mathrm{e}^{\mathrm{i}(q+\xi p).z} \varepsilon(z^0)\delta(z^2) \tilde{S}_1^k(p\,.\,z) s^\beta$$

$$= 4\pi^2 \int_{-\infty}^{\infty} \mathrm{d}\xi (q+\xi p)^\alpha \delta(q^2 + 2\xi q\,.\,p) S_1^k(\xi) s^\beta$$

$$= \frac{2\pi^2}{M\nu} (q+wp)^\alpha S_1^k(w) s^\beta, \tag{6.38}$$

where $S_1^k(w)$ is defined as the Fourier transform of $\tilde{S}_1^k(p\,.\,z)$ in the way $S^k(\alpha)$ was defined from $\tilde{S}^k(z\,.\,p)$ in eqns (6.7). Comparing eqns (6.36), (6.37), and (6.38) with eqn (6.31), we see that since the latter does not have any term of the type $\varepsilon_{\mu\nu\alpha\beta}p^\alpha s^\beta$, the $p^\alpha s^\beta$ term in $I_1^{\alpha\beta}$ has to cancel with a non-leading term, owing to current conservation.

(2)
$$I_2^{\alpha\beta} = \int \mathrm{d}^4z\, \mathrm{e}^{\mathrm{i}q.z}\partial^\alpha\{\varepsilon(z^0)\delta(z^2)\}s\,.\,z\tilde{S}_2^k(p\,.\,z)p^\beta$$
$$= -\mathrm{i}\int_{-\infty}^{\infty}\mathrm{d}\xi\,(q+\xi p)^\alpha p^\beta \int \mathrm{d}^4z\,\mathrm{e}^{\mathrm{i}(q+\xi p).z}\varepsilon(z^0)\delta(z^2)s\,.\,zS_2^k(\xi) -$$
$$-s^\alpha p^\beta \int_{-\infty}^{\infty}\mathrm{d}\xi \int \mathrm{d}^4z\,\mathrm{e}^{\mathrm{i}(q+\xi p).z}\varepsilon(z^0)\delta(z^2)S_2^k(\xi).$$

The second term in the above expression for $I^2_{\alpha\beta}$ will again be cancelled by a non-leading term. Thus we can *effectively* write

$$I_2^{\alpha\beta} \to \mathrm{i}s^\rho \int_{-\infty}^{\infty}\mathrm{d}\xi\, p^\beta(q+\xi p)^\alpha(-\mathrm{i})\frac{\partial}{\partial R_\rho}\left\{-\frac{4\pi^2}{\mathrm{i}}\delta(R^2)\varepsilon(R^0)\right\}_{R=q+\xi p} S_2^k(\xi)$$
$$= -8\pi^2\mathrm{i}\int_{-\infty}^{\infty}\mathrm{d}\xi\, p^\beta(q+\xi p)^\alpha(q+\xi p)^\rho\delta'(q^2+2\xi q\,.\,p)s_\rho S_2^k(\xi)$$
$$= \frac{2\pi^2\mathrm{i}}{(q\,.\,p)^2}p^\beta(q+wp)^\alpha s\,.\,qS_2'^k(w),$$

where $S_2'^k(w) = (\mathrm{d}/\mathrm{d}w)S_2^k(w)$ and we have used the results $s\,.\,p = 0$ in the last step. Thus, finally,

$$I_2^{\alpha\beta} \to \frac{2\pi^2\mathrm{i}}{M^2\nu^2}q^\alpha p^\beta s\,.\,qS_2'^k(w), \tag{6.39}$$

where once again the term proportional to $p^\alpha p^\beta$ has been dropped by comparison with eqn (6.31) and in deference to current conservation. Substituting the expressions of eqns (6.38) and (6.39) in $S^{\mathrm{e}}_{\mu\nu}$, we obtain via the use of eqns (6.37) and (6.36) that

$$\lim_{\mathrm{LC}} S^{\mathrm{e}}_{\mu\nu} = -\frac{1}{4\pi^2}\mathrm{i}\varepsilon_{\mu\nu\alpha\beta}d^{ijk}\left\{\frac{2\pi^2}{M\nu}q^\alpha s^\beta S_1^k(w) - \frac{2\pi^2}{M^2\nu^2}s\,.\,qq^\alpha p^\beta S_2'^k(w)\right\}, \tag{6.40}$$

where we have continued using $i = j = (3)+1/\sqrt{3}(8)$. Identifying the coefficients in the right-hand side of eqn (6.40) with the structure functions $X_1$, $X_2$ of eqn (6.31), we obtain

$$\lim_{\text{Bj}} \nu X_1^{\text{e}}(Q^2, \nu) = G_1^{\text{e}}(w) = -\tfrac{1}{2}d^{ijk}S_1^k(w), \tag{6.41a}$$

$$\lim_{\text{Bj}} M^{-1}\nu^2 X_2^{\text{e}}(Q^2, \nu) = G_2^{\text{e}}(w) = \tfrac{1}{2}d^{ijk}S_2'^k(w). \tag{6.41b}$$

Putting in the appropriate values of $i$, $j$, $k$, etc. in eqns (6.41), we obtain the spin-dependent scale functions for deep inelastic eN scattering to be

$$G_1^{\text{e}}(w) = -\frac{1}{3}\sqrt{\left(\frac{2}{3}\right)}S_1^0(w)-\frac{1}{3}S_1^3(w)-\frac{1}{3\sqrt{3}}S_1^8(w), \tag{6.42a}$$

$$G_2^{\text{e}}(w) = \frac{2}{3}\sqrt{\left(\frac{2}{3}\right)}S_2'^0(w)+\frac{1}{3}S_2'^3(w)+\frac{1}{3\sqrt{3}}S_2'^8(w). \tag{6.42b}$$

Eqns (6.42) are our principal results. From these we can obtain sum rules in the following way. First note the vanishing of an integral

$$\int_0^1 \mathrm{d}w\, G_2^{\text{e}}(w) = \tfrac{1}{2}d^{ijk}\{S_2^k(1)-S_2^k(0)\} = 0. \tag{6.43}$$

In eqn (6.43), $S_2^k(1)$ is zero owing to threshold, and $S_2^k(0)$ vanishes owing to $S_2^k(w)$ being an odd function of $w$. Eqn (6.43) is an interesting sum rule, but we can obtain one that is more interesting as follows. Observe that

$$-\int_{-1}^{1} \mathrm{d}w\, S_1^k(w) = \tilde{S}_1^k(z=0) \equiv s_1^k, \tag{6.44}$$

where

$$s_1^k s_\mu + \cdots = (2\pi)^3\frac{p_0}{M}\langle \text{N}(p), s|J_{\mu5}^{\oplus,k}(0,0)|\text{N}(p), s\rangle$$

$$= (2\pi)^3\frac{p_0}{M}\langle \text{N}(p), s|J_{\mu5}^{k}(0)|\text{N}(p), s\rangle. \tag{6.45}$$

In eqn (6.45) the dots stand for the non-polarization part. Introduce the standard weak axial-vector coupling constant of the nucleon through the relation (Marshak *et al.* 1969)

$$(2\pi)^3\frac{p_0}{M}\langle \text{p}(p), s|J_{\mu5}^{1+i2}(0)|s, \text{n}(p')\rangle \equiv g_A\bar{u}_{\text{p}}(p)\gamma_\mu\gamma_5 u_{\text{n}}(p'). \tag{6.46}$$

The neutron state $|\mathrm{n}\rangle$ is obtained when a charge-lowering operator acts on the proton state $|\mathrm{p}\rangle$:

$$|\mathrm{n}\rangle = Q^{1-\mathrm{i}2}|\mathrm{p}\rangle.$$

Now, since the matrix element $\langle p, s|Q^{1-\mathrm{i}2}J_{\mu}^{1-\mathrm{i}2}(0)|p, s\rangle$ vanishes, the left-hand side of eqn (6.46) may be written as

$$(2\pi)^3 \frac{p_0}{M}\langle p, s|[J_{\mu 5}^{1+\mathrm{i}2}, Q^{1-\mathrm{i}2}]|p', s\rangle$$

$$= 2(2\pi)^3 \frac{p_0}{M}\langle p, s|J_{\mu 5}^{3}(0)|p', s\rangle,$$

so that we obtain the result

$$(2\pi)^3 \frac{p_0}{M}\langle p, s|J_{\mu 5}^{3}(0)|p, s\rangle = \tfrac{1}{2}g_A \bar{u}\gamma_\mu\gamma_5 u. \tag{6.47}$$

The polarization part of the left-hand side of eqn (6.47) is

$$s_{1\mathrm{p}}^{3} s_\mu = \frac{g_A}{2}\frac{1}{2M}\,\mathrm{Tr}\,\tfrac{1}{2}(1+\not{s}\gamma_5)(\not{p}+M)\gamma_\mu\gamma_5 = -\tfrac{1}{2}g_A s_\mu .$$

Comparing this result with eqns (6.45) and (6.46), we have

$$-s_{1\mathrm{p}}^{3} = \tfrac{1}{2}g_A . \tag{6.48}$$

Eqns (6.42), (6.44), and (6.48) lead to the sum rule

$$\int_{-1}^{1} \mathrm{d}w\,\{G_1^{\mathrm{ep}}(w) - G_1^{\mathrm{en}}(w)\} = -\tfrac{2}{3}s_{1\mathrm{p}}^{3} = \tfrac{1}{3}g_A$$

or, because of the evenness of the integrand, to the relation

$$\int_{0}^{1} \mathrm{d}w\,\{G_1^{\mathrm{ep}}(w) - G_1^{\mathrm{en}}(w)\} = \tfrac{1}{6}g_A . \tag{6.49}$$

Recall from eqn (6.32) that

$$\lim_{\varepsilon \gg \nu} P = -\frac{q^2}{\varepsilon}\frac{X_1^{\mathrm{e}} + \nu X_2^{\mathrm{e}}}{\nu W_2^{\mathrm{e}}}$$

and from eqn (6.43) that

$$\int_{0}^{1} \mathrm{d}w\, G_2(w) = 0.$$

The use of this information in eqn (6.49) leads to the Bjorken backward sum rule

$$\lim_{\substack{\varepsilon(\gg\nu)\to\infty \\ Q^2\,\text{large}}} \varepsilon \int_{-q^2/2}^{\infty} \frac{\mathrm{d}\nu}{\nu^2}(P^{\text{ep}}W_2^{\text{ep}} - P^{\text{en}}W_2^{\text{en}}) = -\tfrac{1}{3}g_A. \tag{6.50}$$

## Other applications

*Disconnected parts*

So far we have considered only the connected parts of the light-cone current commutators and omitted the disconnected $c$-number parts. The former are relevant to deep inelastic $l$N scattering. The latter are useful too; they find application in high-energy $e^+e^-$ annihilation into hadrons. Recall from our discussions in Chapter 4, in particular eqns (4.63) and (4.64), that the cross-section for that reaction—in the single-photon exchange approximation—may be written as

$$\sigma_{\text{tot}}^{e^+e^-}(q^2) = -\frac{16}{3}\frac{\pi^2\alpha^2}{q^4}\int \mathrm{d}^4x\, \mathrm{e}^{\mathrm{i}q.x}\langle 0|[J_\mu^{\text{EM}}(x), J^{\mu,\text{EM}}(0)]|0\rangle. \tag{6.51}$$

In the limit when $q_0 \to \infty$ in the frame $\mathbf{q} = 0$, the region with $x^0$ tending to zero is picked up in the commutator. In order to preserve micro-causality, $\mathbf{x}$ must then approach zero also in the contributing region. Thus only the tip of the light-cone contributes to the commutator in the integrand of eqn (6.51). However, we may approach this tip along the surface of the light-cone and thereby use light-cone current algebra. The use of the latter here is equivalent to exploiting the free quark-field structure of the electromagnetic current

$$J_\mu^{\text{EM}} \sim :\bar{q}\gamma_\mu \frac{\lambda^Q}{2} q:, \qquad \lambda^Q = \lambda^3 + \frac{1}{\sqrt{3}}\lambda^8,$$

and employing Wick's theorem. This procedure leads (see Exercise 6.5) to the result (Frishman 1972)

$$\begin{aligned}\langle 0|[J_\mu^{\text{EM}}(x), J_\nu^{\text{EM}}(0)]|0\rangle &\\ \triangleq \sum_i \mathrm{Tr}(\gamma_\mu\gamma^\alpha\lambda^Q\gamma_\nu\gamma^\beta\lambda^Q)&[\{\partial_\alpha\Delta^{(-)}(x)\}\{\partial_\beta\Delta^{(-)}(x)\} - \{\partial_\alpha\Delta^{(+)}(x)\}\{\partial_\beta\Delta^{(+)}(x)\}]\\ = \sum_i Q_i^2 \frac{\mathrm{i}}{3\pi^3}&\varepsilon(x^0)\delta'''(x^2)(g_{\mu\nu}x^2 - 2x_\mu x_\nu).\end{aligned} \tag{6.52}$$

In eqn (6.52) the trace is to be taken both for the gamma and the lambda matrices, $Q_i$ stands for the charge of the $i$th quark, and the summation over $i$

refers to the need for considering all types of quarks present. The substitution of this equation into eqn (6.51) leads to the relation

$$\lim_{q^2\to\infty} \sigma_{\text{tot}}^{e^+e^-}(q^2) = \frac{4\pi\alpha^2}{3q^2}\sum_i Q_i^2 = \lim_{q^2\to\infty} \sigma_{e^+e^-\underset{\gamma^*}{\to}\mu^+\mu^-}(q^2)\sum_i Q_i^2. \tag{6.53}$$

If there is only one type of triplet and the quarks carry fractional charges in the manner prescribed by Gell-Mann and Zweig (Kokkedee 1969), the factor $\sum_i Q_i^2$ becomes $\frac{2}{3}$—a number not in good agreement with present experiment (Little 1972). If the quarks are taken to be 'charmed' objects of the Han–Nambu type, with integral charges, $\sum_i Q_i^2$ becomes 4 when charmed hadrons are produced and 2 when they are not (see Chapter 1, p. 41). Another approach (Gell-Mann 1972) is to retain the fractionally-charged triplet but add a new SU(3) quantum number called colour. There are three colours (dubbed red, blue, and white by Gell-Mann) and there is a colour-SU(3), but all physical hadrons are supposedly singlets in that group. (This scheme is then equivalent to that of GMZ quarks with parastatistics of rank 3.) Thus, for instance, the $q\bar{q}$ configuration for mesons is $(1/\sqrt{3})(q_R\bar{q}_R+q_B\bar{q}_B+q_W\bar{q}_W)$ and the $qqq$ configuration for baryons is $(1/\sqrt{6})(q_Rq_Bq_W-q_Bq_Rq_W+q_Wq_Rq_B-q_Rq_Wq_B+q_Bq_Wq_R-q_Wq_Bq_R)$. In this picture the factor $\sum_i Q_i^2$ becomes 2, which turns out to be closer to experiment than the earlier number $\frac{2}{3}$. (However, according to the latest experiments $\sigma_{\text{tot}}^{e^+e^-}(q^2)/\sigma_{e^+e^-\to\mu^+\mu^-}(q^2)$ rises almost linearly with $q^2$ up to $q^2 = 25\ \text{GeV}^2$ beyond a numerical value of 5, and the situation does look somewhat unsettling.) The need for colour or parastatistics of rank 3 is also felt in connection with hadron spectroscopy based on quarks, as well as with the theoretical understanding of the $\pi^0 \to 2\gamma$ decay problem (Gell-Mann 1972), but these topics are beyond the scope of the present book.

*Application of the bilocal algebra to two-current reactions*

We shall now take up situations which involve applications of the closed algebra of the bilocal currents of eqns (5.44) and (5.45). Consider an inelastic reaction induced by an initial current and generating a final current, as illustrated in Fig. 40. Here $(q_1+p)^2 = s$, $(q_1-q_2)^2 = t$, and $(q_2-p)^2 = u$. Such a diagram is relevant to the Compton-type terms in reactions such as $e+N \to e+\mu^++\mu^-+$'anything', $e+N\to e+W^\pm+$'anything' with weak interactions and electromagnetism treated to the lowest possible order. We

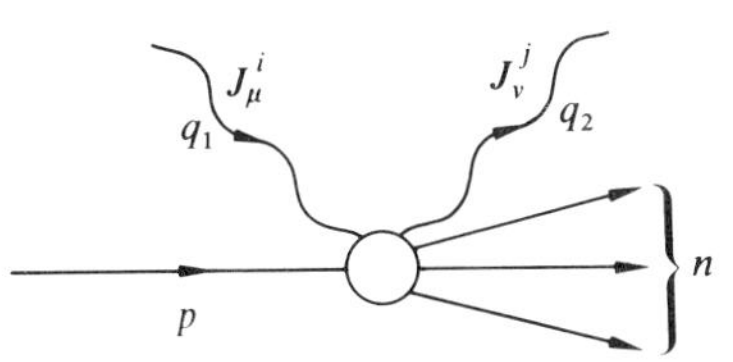


FIG. 40. Two-current inelastic interactions.

shall show that the bilocal algebra can be applied to this process, provided certain kinematic conditions are satisfied. The relevant amplitude is

$$\mathcal{M}^{n}_{\mu\nu} = \int \mathrm{d}^4x \exp\left(-\mathrm{i}\frac{q_1+q_2}{2}.x\right)\left\langle n \left| T^* J^i_\mu\left(\frac{x}{2}\right) J^j_\nu\left(-\frac{x}{2}\right)\right| \mathrm{N}\right\rangle, \tag{6.54}$$

where $T^*$ stands for the physical ordered product and $J^i_\mu$ is a hadronic current in the chiral symmetric group $\mathrm{SU}(3)\times\mathrm{SU}(3)$. (The reader is reminded that a physical product differs from a regular $T$ product by some tensors with real polynomial coefficients known as seagulls.) In order to calculate cross-sections for relevant reactions we need to consider quantities such as

$$\sum_n (2\pi)^3\delta^{(4)}(p+q_1-q_2-p_n)\mathcal{M}^{*n}_{\nu\beta}\mathcal{M}^{n}_{\mu\alpha}.$$

Thus the object of ultimate interest is the tensor

$$\begin{aligned} T^{ijkl}_{\beta\nu\mu\alpha} = (2\pi)^3\frac{p_0}{M}\int \mathrm{d}^4x \int \mathrm{d}^4y \int \mathrm{d}^4z\, \mathrm{e}^{-\mathrm{i}R.x+\mathrm{i}R.y-\mathrm{i}S.z} \times \\ \times \left\langle \mathrm{N}(p)\left| \overline{T}^*\left\{ J^i_\beta\left(\frac{z-y}{2}\right) J^j_\nu\left(\frac{y+z}{2}\right)\right\}\times\right.\right. \\ \left.\left. \times\, T^*\left\{ J^i_\mu\left(\frac{x-z}{2}\right) J^l_\alpha\left(-\frac{x+z}{2}\right)\right\}\right| \mathrm{N}(p)\right\rangle. \end{aligned} \tag{6.55}$$

In eqn (6.55) $R \equiv \frac{1}{2}(q_1+q_2)$ and $S \equiv q_1-q_2$. With respect to this equation certain comments are in order. First, in order to apply the bilocal algebra, we have to generalize from commutators to physical products. We postulate that the algebra goes through for $T^*$ products also (in the sense of the structure of leading light-cone singularities remaining to be that of free quark-field theory) provided we make the change

$$\Delta(z) \mathrel{\widehat{=}} -\frac{1}{2\pi}\varepsilon(z^0)\delta(z^2) \to \Delta_{\mathrm{F}}(z) \mathrel{\widehat{=}} \frac{1}{2\pi^2}(-z^2+\mathrm{i}\varepsilon)^{-1}$$

as prescribed by free field theory. Such a modification is not necessarily trivial. To give a counterexample, it does not work in the case of Compton scattering from a target of spin-0, where seagull terms occur explicitly (Fig. 41). Second, for the applicability of the bilocal algebra we have to choose the positions $\frac{1}{2}(z-y)$, $\frac{1}{2}(z+y)$, $\frac{1}{2}(x-z)$ and $\frac{1}{2}(-x-z)$ of the four currents in eqn (6.55) to be collinear on a light-like ray. To ensure this, take the variables $|q_1^2|$, $|s|$, $|t|$, $|q_2^2|$ to infinity in such a way that the ratios $q_1^2/s$, $t/q_1^2$, $q_2^2/q_1^2$ are all

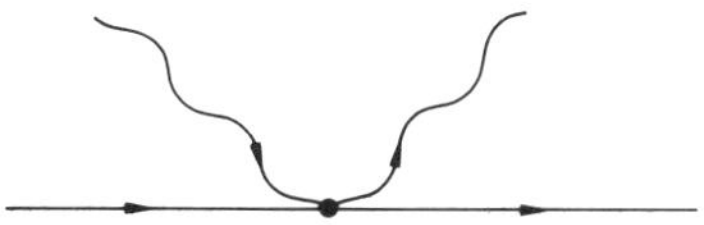


FIG. 41. Seagull term in Compton scattering.

kept finite. In this region $|R^2|$, $|S^2|$, and $|R\,.\,S|$ automatically approach infinity, with the ratio between any two of them staying finite. Then in the integrals of eqn (6.55) we have $x^2 \to 0$, $y^2 \to 0$, and $z^2 \to 0$ by the Riemann–Lebesgue lemma. (For this part of the argument the physical ordered product may be replaced by a retarded commutator, in conformity with the reduction formalism, to enforce micro-causality.) Now write $R^\mu = ue^\mu + a^\mu$, $S^\mu = ve^\mu + b^\mu$, where $e^\mu$ is a light-like vector ($e^2 = 0$). This implies that $R^2 = 2ue\,.\,a + a^2$, $S^2 = 2ve\,.\,a + b^2$ and $R\,.\,S = ve\,.\,a + ve\,.\,b$. We have chosen $a^\mu$, $b^\mu$ to be fixed, time-like vectors and $u$, $v$ to be scalar variables. In the asymptotic region described above, $|u|$, $|v|$ have to go to infinity with the ratio $u/v$ fixed. The phase $S\,.\,z$ in the integral of eqn (6.55) has to be bounded, which implies under the present circumstances that $z\,.\,e = 0$. Thus $z$ is either proportional to $e$ or orthogonal to it. But if $z$ is orthogonal to $e$, it has to be space-like—as is seen immediately by going to a frame where $e = (1, 0, 0, 1)$; then the limit $z^2 \to 0$ will be possible only if $z$ itself tends to zero. Thus, in limit of interest, $z$ must become proportional to $e$ (in the null case the proportionality constant merely vanishes). Similarly, from the boundedness of the phases $R\,.\,x$, $R\,.\,y$, we obtain that $x$ and $y$ are also proportional to $e$ in this limit. Thus $0, x, y, z$ are collinear on a light-like ray. Hence in deep inelastic reactions involving two highly virtual currents, we can find suitable kinematic regions where the bilocal algebra can be applied. We shall omit the details of these applications which have been made to reactions such as $\mathrm{e+N \to e+W+}$'anything' (Jarlskog and Roy 1972) whose differential cross-sections have been related to the deep inelastic lepton–nucleon structure functions.

## Fixed-mass sum rules by light-plane methods

In Chapter 5 we alluded to the class of problems with currents carrying fixed masses, where light-plane commutators proved useful. We shall now discuss the derivation of sum rules for such reactions by light-plane methods (Dicus, Jackiw, and Teplitz 1971). Historically, such sum rules were first derived by the (not always successful) infinite-momentum technique (Adler and Dashen 1969). The pitfalls of that approach may be avoided, however, if we use light-plane current algebra instead. Let us return to eqn (6.27) and generalize it to the scattering from a nucleon of *conserved* vector currents with arbitrary SU(3) indices. Thus we now have

$$W^{ij}_{\mu\nu}+S^{ij}_{\mu\nu}$$

$$= (2\pi)^2\frac{p_0}{M}\int \mathrm{d}^4x\, \mathrm{e}^{\mathrm{i}q.x}\langle \mathrm{N}(p), s|[J^i_\mu(x), J^j_\nu(0)]|\mathrm{N}(p), s\rangle$$

$$= \left(-g^{\mu\nu}+\frac{q^\mu q^\nu}{q^2}\right)W^{ij}_{\mathrm{L}}+\frac{1}{M^2}\left\{p^\mu p^\nu-\frac{p\,.\,q}{q^2}(p^\mu q^\nu+p^\nu q^\mu)+g^{\mu\nu}\frac{(p\,.\,q)^2}{q^2}\right\}W^{ij}_2+$$

$$+\frac{\mathrm{i}}{M}\varepsilon_{\mu\nu\alpha\beta}q^\alpha s^\beta X^{ij}_1+\frac{\mathrm{i}}{M^3}\varepsilon_{\mu\nu\alpha\beta}q^\alpha p^\beta q\,.\,sX^{ij}_2. \tag{6.56}$$

In the last step of eqn (6.56) we have carried out the decomposition into appropriate covariant independent tensors (cf. eqns (I.7), (I.17), and (6.31)). We take the invariant functions of $Q^2$ and $\nu$ to be made up of parts which are symmetric and antisymmetric in $i$ and $j$, labelled $(ij)$ and $[ij]$ respectively,

$$W^{ij}_{\mathrm{L},2} = W^{(ij)}_{\mathrm{L},2}+\mathrm{i}W^{[ij]}_{\mathrm{L},2}, \tag{6.57a}$$

$$X^{ij}_{1,2} = X^{(ij)}_{1,2}+\mathrm{i}X^{[ij]}_{1,2}. \tag{6.57b}$$

The 'crossing' properties of the structure functions are

$$W^{(ij)}_{\mathrm{L},2}(Q^2,\nu) = -W^{(ij)}_{\mathrm{L},2}(Q^2,-\nu), \qquad W^{[ij]}_{\mathrm{L},2}(Q^2,\nu) = W^{[ij]}_{\mathrm{L},2}(Q^2,-\nu),$$
$$X^{(ij)}_1(Q^2,\nu) = -X^{(ij)}_1(Q^2,-\nu), \qquad X^{[ij]}_1(Q^2,\nu) = X^{[ij]}_1(Q^2,-\nu),$$
$$X^{(ij)}_2(Q^2,\nu) = X^{(ij)}_2(Q^2,-\nu), \qquad X^{[ij]}_2(Q^2,\nu) = -X^{[ij]}_2(Q^2,-\nu). \tag{6.58}$$

Hermiticity and invariance under time-reversal are satisfied so long as the invariants occurring in eqn (6.58) are real.

Introducing again the light-plane notation of Chapter 5, we set $q^+=0$ in eqn (6.56), choose $\mu=+$, and integrate both sides over $q^-$. This leads to

$$\frac{M}{2\pi}\int_{-\infty}^{\infty}\frac{\mathrm{d}\nu}{p^+}\{W^{+\nu,ij}(p,q)+S^{+\nu,ij}(p,q)\}$$

$$= (2\pi)^2\frac{p_0}{M}\int_{-\infty}^{\infty}\mathrm{d}x^-\int \mathrm{d}^2x_\perp\,\exp(-\mathrm{i}\mathbf{q}_\perp.\mathbf{x}_\perp)\times$$

$$\times\langle \mathrm{N}|[J^{+,i}(x), J^{\nu,j}(0)]|\mathrm{N}\rangle_{x^+=0}. \tag{6.59}$$

In eqn (6.59) the limit $x^+\to 0$ has been interchanged with the integration over $x^-$, i.e. we have assumed that

$$\left\langle \mathrm{N}\left|\left[\int \mathrm{d}x^-\, J^i_+(x), J^j_\nu(0)\right]\right|\mathrm{N}\right\rangle_{x^+=0} = \int \mathrm{d}x^-\,\langle \mathrm{N}|[J^i_+(x), J^j_\nu(0)]_{x^+=0}|\mathrm{N}\rangle. \tag{6.60}$$

The rationale behind eqn (6.60) will be given later. But note meanwhile that eqn (6.59) is a fixed-mass sum rule since, for $q^+ = 0$, the mass-squared carried by the currents is $-q_\perp^2$, which is fixed in the integration over $q^-$. The right-hand side of the sum rule is given by a light-plane current commutator. More specific information may be secured by taking $\nu = +, -$ and using eqns (5.87a) and (5.87c), now without any SU(3)-breaking terms. (The case with $\nu = r$ then turns out to be redundant.) For this purpose it is convenient to use the symmetric and antisymmetric bilocal currents introduced earlier. The reader is reminded that

$$J_\mu^{\oplus\atop\ominus}(x, 0) = \tfrac{1}{2}\{J_\mu^k(x, 0) \pm J_\mu^k(0, x)\}.$$

Consistent with general symmetry requirements, we can make the following form-factor decompositions of matrix elements of these currents between single-nucleon states:

$$(2\pi)^3 \frac{p_0}{M} \langle \mathrm{N}| J_\mu^{\oplus,k}(x, 0)|\mathrm{N}\rangle$$
$$= \tilde{S}^k(x^2, x\,.\,p)p_\mu + M^2 \tilde{S}_1^k(x^2, x\,.\,p)x_\mu, \tag{6.61a}$$

$$(2\pi)^3 \frac{p_0}{M} \langle \mathrm{N}| J_\mu^{\ominus,k}(x, 0)|\mathrm{N}\rangle$$
$$= \tilde{A}^k(x^2, x\,.\,p)p_\mu + M^2 \tilde{A}_1^k(x^2, x\,.\,p)x_\mu, \tag{6.61b}$$

$$(2\pi)^3 \frac{p_0}{M} \langle \mathrm{N}| J_{\mu 5}^{\oplus,k}(x, 0)|\mathrm{N}\rangle$$
$$= x\,.\,s\tilde{S}_5^k(x^2, x\,.\,p)p_\mu + M^2 x\,.\,s\tilde{S}_{5,1}^k(x^2, x\,.\,p)x_\mu + \tilde{S}_{5,3}(x^2, x\,.\,p)s_\mu, \tag{6.61c}$$

$$(2\pi)^3 \frac{p_0}{M} \langle \mathrm{N}| J_{\mu 5}^{\ominus,k}(x, 0)|\mathrm{N}\rangle$$
$$= x\,.\,s\tilde{A}_5^k(x^2, x\,.\,p)p_\mu + M^2 x\,.\,s\tilde{A}_{5,1}(x^2, x\,.\,p)x_\mu + \tilde{A}_{5,3}(x^2, x\,.\,p)s_\mu. \tag{6.61d}$$

Some comments on these equations are in order. The presence of any explicitly spin-dependent term in eqns (6.61a) and (6.61b) is ruled out by $P$- and $T$-invariance. Moreover, on spin-averaging, the right-hand sides of the last two of the above equations vanish as expected. The quantities $\tilde{S}^k$ and $\tilde{A}^k$ are even and odd functions of $x$ respectively corresponding to the evenness or oddness property of the relevant matrix element. Finally, to establish connections with the corresponding matrix element of the local current $J_\mu^k(0)$ note that $\tilde{S}^k(0, 0) = \Gamma^k/M$, where

$$(2\pi)^3 \frac{p_0}{M} \langle \mathrm{N}| J_\mu^k(0)|\mathrm{N}\rangle = \Gamma^k \frac{p_\mu}{M}. \tag{6.62}$$

If we choose $\nu = +$ in eqn (6.59), make use of eqn (6.56) in the left-hand side, and evaluate the light-plane commutator following eqn (5.87a), we are led to the sum rule

$$\int_{-\infty}^{\infty} \mathrm{d}\nu\, W_2^{ij}(Q^2, \nu) = \mathrm{i} f^{ijk}\Gamma^k,$$

or

$$\int_{0}^{\infty} \mathrm{d}\nu\, W_2^{[ij]}(Q^2, \nu) = \tfrac{1}{2} f^{ijk}\Gamma^k, \tag{6.63}$$

where we have used the relevant 'crossing' information of eqns (6.58). Eqn (6.63) is the celebrated Dashen–Fubini–Gell-Mann sum rule. The next step is to set $\nu = -$ in eqn (6.59). By use of eqns (6.56) and (5.87c) and discarding terms with $q^+$ in the left-hand side and over-all derivatives in the $x^-$ integration, we then obtain (with $Q^2 = q_\perp^2$)

$$M \int_{-\infty}^{\infty} \frac{\mathrm{d}\nu}{p^+} \left\{ -W_{\mathrm{L}}^{ij} + \frac{1}{M^2}\left(p^+p^- + \frac{M\nu}{Q^2}\mathbf{p}_\perp \cdot \mathbf{q}_\perp\right) W_2^{ij} + \frac{\mathrm{i}}{M}\varepsilon_{rs} q^r s^s X_1^{ij} + \right.$$
$$\left. + \frac{\mathrm{i}}{M^3}\varepsilon_{rs} q^r p^s (q^- s^+ - \mathbf{q}_\perp \cdot \mathbf{s}_\perp) X_2^{ij} \right\}$$
$$= (2\pi)^2 \frac{p_0}{M} \langle \mathrm{N}| \mathrm{i} f^{ijk} J^{-,k}(0) - \tfrac{1}{4} f^{ijk} q^r \int_{-\infty}^{\infty} \mathrm{d}x^-\, \varepsilon(x^-) J_r^{\oplus,k}(x, 0) -$$
$$- \frac{\mathrm{i}}{4} f^{ijk} q_r \varepsilon^{rs} \int_{-\infty}^{\infty} \mathrm{d}x^-\, \varepsilon(x^-) J_{s5}^{\ominus,k}(x, 0) +$$
$$+ \frac{\mathrm{i}}{4} d^{ijk} q^r \int_{-\infty}^{\infty} \mathrm{d}x^-\, \varepsilon(x^-) J_r^{\ominus,k}(x, 0) -$$
$$- \frac{1}{4} d^{ijk} q_r \varepsilon^{rs} \int_{-\infty}^{\infty} \mathrm{d}x^-\, \varepsilon(x^-) J_{s5}^{\ominus,k}(x, 0) |\mathrm{N}\rangle_{x^+ = 0 = \mathbf{x}_\perp}. \tag{6.64}$$

Eqn (6.64) may be simplified considerably. On the left-hand side we may use the oddness or evenness properties of $W_{\mathrm{L},2}(Q^2 = q_\perp^2, \nu)$, $X_{1,2}(Q^2 = q_\perp^2, \nu)$ described in eqns (6.57) and (6.58). On the right-hand side the symmetry or antisymmetry of the matrix elements may be exploited, leading immediately to the elimination of the contributions from the symmetric bilocal currents. These operations yield the following result:

$$\frac{2iM}{p^+}\int_0^\infty d\nu\left(-W_L^{[ij]}+\frac{p^+p^-}{M^2}W_2^{[ij]}\right)+\frac{2\mathbf{p}_\perp\cdot\mathbf{q}_\perp}{Q^2p^+}\int_0^\infty d\nu\,\nu W_2^{(ij)}-$$

$$-\frac{2\varepsilon^{rs}q_s}{p^+}\int_0^\infty d\nu\left\{s_rX_1^{[ij]}+\frac{s^+p_r\nu}{Mp^+}X_2^{[ij]}\right\}+\frac{2i\varepsilon^{rs}p_rq_s}{p^+M^2}\times$$

$$\times\left(\mathbf{p}_\perp\cdot\mathbf{q}_\perp\frac{s^+}{p^+}-\mathbf{q}_\perp\cdot\mathbf{s}_\perp\right)\int_0^\infty d\nu\,X_2^{(ij)}$$

$$=(2\pi)^3\frac{p_0}{M}\left\langle N\left|if^{ijk}\left\{J^{-,k}(0)-\frac{1}{2}q_r\varepsilon^{rs}\int_0^\infty dx^-\,J_{s5}^{\ominus,k}(x,0)\right\}+\right.\right.$$

$$\left.\left.+\frac{i}{2}d^{ijk}q^r\int_0^\infty dx^-J_r^{\ominus,k}(x,0)\right|N\right\rangle_{x^+=0=\mathbf{x}_\perp}. \qquad (6.65)$$

The terms in eqn (6.65) which are symmetric in $i, j$ may be separated from those antisymmetric in the same. Similarly, the spin-dependent structure functions $X_{1,2}$ may be separated from the spin-independent ones $W_{L,2}$ by making use of eqns (6.61) and (6.62) on the right-hand side of eqn (6.65). We are thus led to four relations, given below.

$$\frac{2iM}{p^+}\int_0^\infty d\nu\left\{-W_L^{[ij]}(Q^2,\nu)+\frac{p^+p^-}{M^2}W_2^{[ij]}(Q^2,\nu)\right\}=if^{ijk}\Gamma^k\frac{p^-}{M},$$

$$\frac{2\mathbf{p}_\perp\cdot\mathbf{q}_\perp}{Q^2p^+}\int_0^\infty d\nu\,\nu W_2^{(ij)}=\frac{i}{2}d^{ijk}q^rp_r\int_0^\infty dx^-\,\tilde{A}^k(0,p^+x^-),$$

$$-\frac{2\varepsilon^{rs}q_s}{p^+}\int_0^\infty d\nu\left(s_rX_1^{[ij]}+\frac{s^+p_r\nu}{Mp^+}X_2^{[ij]}\right)$$

$$=-\frac{i}{2}f^{ijk}q_r\varepsilon^{rs}\int_0^\infty dx^-\{x^-s^+p_s\tilde{A}_5^k(0,p^+x^-)+s_s\tilde{A}_{5,3}^k(0,p^+x^-)\},$$

$$\frac{2i\varepsilon^{rs}p_rq_s}{p^+M^2}\left(\mathbf{p}_\perp\cdot\mathbf{q}_\perp\frac{s^+}{p^+}-\mathbf{q}_\perp\cdot\mathbf{s}_\perp\right)\int_0^\infty d\nu\,X_2^{(ij)}=0.$$

Algebraic simplifications of the above equations, as well as the use of eqn (6.63), lead to the following independent and Lorentz-invariant sum rules

$$\int_0^\infty d\nu\, W_{\rm L}^{[ij]}(Q^2,\nu) = 0, \tag{6.66a}$$

$$\int_0^\infty d\nu \frac{\nu}{Q^2} W_2^{(ij)}(Q^2,\nu) = -\frac{i}{4} d^{ijk} \int_0^\infty d\alpha\, \tilde{A}^k(0,\alpha), \tag{6.66b}$$

$$\int_0^\infty d\nu\, X_1^{[ij]}(Q^2,\nu) = -\frac{i}{4} f^{ijk} \int_0^\infty d\alpha\, \tilde{A}_{5,3}^k(0,\alpha), \tag{6.66c}$$

$$\int_0^\infty d\nu \frac{\nu}{M} X_2^{[ij]}(Q^2,\nu) = -\frac{i}{4} f^{ijk} \int_0^\infty d\alpha\, \tilde{A}_5^k(0,\alpha), \tag{6.66d}$$

$$\int_0^\infty d\nu\, X_2^{(ij)}(Q^2,\nu) = 0. \tag{6.66e}$$

Eqns (6.66), in addition to eqns (6.63), constitute the fixed-mass sum rules in inelastic lepton–hadron reactions, which are immediately obtainable from light-plane current algebra. *The remarkable feature of these equations is that the integrals concerned are independent of* $Q^2$. By taking $Q^2 \to \infty$, we may rewrite them for the Bjorken scaling region in terms of the structure functions $F_{\rm L} \equiv \lim 2MwW_{\rm L}$, $F_2 \equiv \lim \nu W_2$, $G_1 \equiv \lim \nu X_1$, $G_2 \equiv \lim M^{-1} \times \nu^2 X_2$:

$$\int_0^\infty d\nu\, W_2^{[ij]}(Q^2,\nu) = \int_0^\infty \frac{dw}{w} F_2^{[ij]}(w) = \tfrac{1}{2} f^{ijk} \Gamma^k, \tag{6.67a}$$

$$\int_0^\infty d\nu\, W_{\rm L}^{[ij]}(Q^2,\nu) = \frac{Q^2}{4M^2} \int_0^\infty \frac{dw}{w^3} F_{\rm L}^{[ij]}(w) = 0, \tag{6.67b}$$

$$\int_0^\infty d\nu\, W_2^{(ij)}(Q^2,\nu) \frac{\nu}{Q^2} = \int_0^1 \frac{dw}{2Mw^2} F_2^{(ij)}(w), \tag{6.67c}$$

$$\int_0^\infty d\nu\, X_1^{[ij]}(Q^2,\nu) = \int_0^1 \frac{dw}{w} G_1^{[ij]}(w), \tag{6.67d}$$

$$\int_0^\infty d\nu\, \nu X_2^{[ij]}(Q^2,\nu) = \int_0^1 \frac{dw}{w} M G_2^{[ij]}(w), \tag{6.67e}$$

$$\int_0^\infty \mathrm{d}\nu\, X_2^{(ij)}(Q^2, \nu) = \frac{Q^2}{2M^2}\int_0^1 \mathrm{d}\omega\, G_2^{(ij)}(w) = 0. \tag{6.67f}$$

Eqn (6.67c) is called the Cornwall–Corrigan–Norton (Cornwall, Corrigan, and Norton 1971) sum rule and eqn (6.67f) is the Burkhardt–Cottingham sum rule (Burkhardt and Cottingham 1970). It should be noted that the old infinite-momentum approach gave incorrect answers for eqns (6.67c), (6.67d), and (6.67e) with zeros in their right-hand sides. This error was a consequence of the improper handling of Z-('class 2') diagrams (Adler and Dashen 1969) in that method. In Regge language, fixed poles in the current–nucleon scattering amplitude—of the form of real polynomials—are the sources of mischief in these instances. The fact that the light-plane approach is able to overcome this difficulty is evidence of its superiority, at least in this respect.

We now clarify the assumed interchange of limit and integration in eqn (6.60) and return to eqns (6.56) and (6.59), which were central to our above considerations. With all indices suppressed the equations are of the form

$$W(Q^2, \nu)+S(Q^2, \nu) = \int \mathrm{d}^4x\, \mathrm{e}^{iq.x}\langle \mathrm{N}\, [J_1(x), J_2(0)]|\mathrm{N}\rangle(2\pi)^2\frac{p_0}{M}, \tag{6.68}$$

where $J_1$ and $J_2$ are two appropriate currents. First set $q^+$ to zero in eqn (6.68), and then integrate both sides over $q^-$. This procedure yields

$$\frac{M}{2\pi}\int_{-\infty}^{\infty}\frac{\mathrm{d}\nu}{p^+}\{W(q_\perp^2, \nu)+S(q_\perp^2, \nu)\}$$
$$= (2\pi)^2\frac{p_0}{M}\int \mathrm{d}^2x_\perp \exp(-\mathrm{i}\mathbf{q}_\perp . \mathbf{x}_\perp)\times$$
$$\times\left\langle \mathrm{N}\left|\left[\int_{-\infty}^{\infty}\mathrm{d}x^-\, J_1(x), J_2(x)\right]_{x^+=0}\right|\mathrm{N}\right\rangle. \tag{6.69}$$

Now consider the full current-scattering amplitude $T$, of which $W+S$ is only the absorptive part,

$$T(Q^2, \nu) = (2\pi)^2\frac{p_0}{M}2\mathrm{i}\int \mathrm{d}^4x\, \mathrm{e}^{iq.x}\langle \mathrm{N}|T^*J_1(x)J_2(0)|\mathrm{N}\rangle. \tag{6.70}$$

The covariant physical ordered ($T^*$) product on the right-hand side of eqn (6.70) is equal to the $T_+$ product, ordered along the $x^+$ direction, modulo 'seagull' terms which are singular, real, and non-vanishing only at the tip of the light cone. This follows from the fact that a step function in time differs from a step function in the light-plane variable $x^+$ only for space-like separations, where ordering is rendered immaterial by causality. The seagull terms, when Fourier-transformed, only generate real polynomials in $q$ and cannot contribute to a commutator which is the discontinuity of an ordered product across cuts on the real axis of the complex $\nu$-plane. Thus

$$T(Q^2, \nu) = T_+(Q^2, \nu) + \text{`seagull' terms},$$

where

$$T_+(Q^2, \nu) = (2\pi)^2 p_0 M^{-1} 2\mathrm{i} \int \mathrm{d}^4x\, \mathrm{e}^{\mathrm{i}q.x} \langle \mathrm{N}| T_+ J_1(x) J_2(0) |\mathrm{N}\rangle.$$

Consider the limit $q^- \to \infty$ with the other components of $q$ held fixed. Then, after integration by parts, we obtain

$$\begin{aligned}\lim T_+ &= -\lim \frac{2}{q^-} \int \mathrm{d}^4x\, \mathrm{e}^{\mathrm{i}q.x} \partial_+ \langle \mathrm{N}| T_+ J_1(x) J_2(0) |\mathrm{N}\rangle (2\pi)^2 \frac{p_0}{M} \\ &= -\frac{2}{q^-}(2\pi)^2 \frac{p_0}{M} \int \mathrm{d}^2x_\perp \int_{-\infty}^{\infty} \mathrm{d}x^- \, \exp(\mathrm{i}q^+x^-) \exp(-\mathrm{i}\mathbf{q}_\perp \cdot \mathbf{x}_\perp) \times \\ &\qquad \times \langle \mathrm{N}| [J_1(x), J_2(x)]_{x^+=0} |\mathrm{N}\rangle + O\left(\frac{1}{(q^-)^2}\right).\end{aligned}$$

Thus we have

$$\begin{aligned}\lim_{q^- \to \infty} T = -\frac{2}{q^-} \int \mathrm{d}^2x_\perp \int_{-\infty}^{\infty} \mathrm{d}x^- \, \exp(\mathrm{i}q^+x^-) \exp(-\mathrm{i}\mathbf{q}_\perp \cdot \mathbf{x}_\perp) \times \\ \times \langle \mathrm{N}| [J_1(x), J_2(0)] |_{x^+=0} |\mathrm{N}\rangle + \\ + \text{real polynomials in } q^- + O\left(\frac{1}{(q^-)^2}\right).\end{aligned} \tag{6.71}$$

A comparison of eqn (6.71) with eqn (6.69) shows that the latter involves $\lim_{q^- \to \infty} q^- T(q^+ = 0)$, whereas the interchanged order as in eqn (6.60) evaluates $(\lim_{q^- \to \infty} q^- T)_{q^+=0}$. Suppose the amplitude $T$ contains a pole term generated by the external current, i.e.

$$T(Q^2, \nu) \simeq \frac{1}{q^2 - \mu^2} R(\nu) + \cdots.$$

Then

$$\left(\lim_{q^- \to \infty} q^- T\right)_{q^+=0} \simeq P \frac{1}{2q^+} R(\infty)\bigg|_{q^+=0} = 0, \tag{6.72}$$

where we have assumed that $R(\infty)$ is finite, i.e. that the commutator exists. On the other hand,

$$\lim_{q^- \to \infty} q^- T(q^+ = 0) \simeq - \lim_{\nu \to \infty} \frac{M\nu R(\nu)}{p^+(q_\perp^2 + \mu^2)}. \tag{6.73}$$

From eqns (6.72) and (6.73), we extract the condition for the validity of the interchange of eqn (6.69),

$$\lim_{\nu \to \infty} \nu R(\nu) = 0. \tag{6.74}$$

When the consequent sum rule converges, there is hope for the validity of eqn (6.74), since convergence implies the vanishing of the discontinuity of $\nu R(\nu)$ in the limit of large $\nu$. On the other hand, if eqn (6.74) is not true, $\int_{-\infty}^{\infty} dx^- J_1(x)$ (or more exactly $\int dx^- J^{+,i}(x)$) is not a local operator in $\mathbf{x}_\perp$, i.e. its commutator with another local operator at the origin does not vanish for $x^+ = 0$, $\mathbf{x}_\perp \neq 0$. This is demonstrated by eqn (6.73), when coupled to eqn (6.71), in that the commutator has a non-polynomial $q_\perp^2$ dependence. This strange behaviour of $\int_{-\infty}^{\infty} dx^- J^{+,i}(x)$ however, would be somewhat peculiar, since it is known that the charge $\int d^2x_\perp \int_{-\infty}^{\infty} dx^- J^{+,i}(x)$ is a well-behaved physical operator independent of $x^+$. These considerations suggest that the assumption of eqn (6.74) is probably sufficiently weak for most purposes of interest.

## Exercises

6.1. Derive eqn (6.13) establishing the relation between $F_1^{ij}(w)$ and $F_2^{ij}(w)$.

6.2. Show from light-cone current commutators that the ratio $R^e \equiv \sigma_L^e/\sigma_T^e$ in inelastic electron scattering vanishes in the Bjorken limit as $\nu^{-1}$ times a function of $w$.

6.3. Using the form of the four-vector

$$s^\mu = \frac{1}{M}\left(\mathbf{p}\cdot\hat{\mathbf{n}}, M\hat{\mathbf{n}} + \frac{\mathbf{p}\cdot\hat{\mathbf{n}}}{p_0 + M}\mathbf{p}\right),$$

obtain two independent relations among the four tensors listed in eqns (6.30) which may be used to eliminate the last two of them and obtain eqn (6.31).

6.4. Derive eqn (6.52) for the disconnected part of the light-cone commutator of two electromagnetic currents and verify eqn (6.53).

# CONCLUSION

THE theoretical ideas on lepton–hadron processes considered in this book possess an underlying unity. This is suggested by the similarity of results, such as Bjorken scaling and sum rules, obtained using one approach or the other. There are, moreover, conceptual links existing between nearly massless partons responding to large momentum transfers and scale invariance at short and light-like distances. They connect the point-like and incoherent behaviour of partons with the free-field nature of leading light-cone singularities in products of currents. They relate the presence of infinite towers of operators with canonical asymptotic dimensions to the existence of a field theory that is asymptotically free. We can make a more precise statement on this unity. There is a dualism (Polkinghorne 1972; Nash 1972; Roy 1972) between the parton and light-cone approaches with scale invariance, especially of free massless field theory, forming the critical link. The theme of a dualism is of course a recurrent one in physics, perhaps because most physicists subscribe to the philosophy of the unity of opposites. The dualism between the parton and light-cone descriptions of deep inelastic scattering is of interest in this context, since they developed from considerably different attitudes. Although both emphasize the notion of effectively free constituents, the parton picture is concerned with their presence in the nucleon (cf. $|\mathrm{N}\rangle \to \sum_l a_l|l\rangle$), whereas the light-cone approach focuses on the structure of hadronic currents in terms of them (cf. $J^i_\mu \sim :\bar{q}\gamma_\mu(\lambda^i/2)q:$). The parton scheme has a rough and crude appearance but the merit of flexibility and improvizability (Feynman 1972). Light-cone physics, in contrast, is based on precisely stated, mathematically elegant assumptions, but is restricted in regard to applications by the burden of a formal structure (Frishman 1972). However, the equivalence between the two descriptions for the case of deep inelastic lepton–nucleon scattering $l+\mathrm{N} \to l'+$'anything' is clear and complete, as demonstrated below.

We consider eN scattering for simplicity, but the argument holds as well for similar reactions induced by neutrinos and antineutrinos. The parton-model expression for the tensor of interest is

$$\lim_{\mathrm{Bj}} W^{\mathrm{e}}_{\mu\nu} = (2\pi)^2 \frac{p_0}{M} \sum_l P(l) \int_0^1 \mathrm{d}\eta \sum_i n_{li} f_{li}(\eta) Q_i^2 \times$$

$$\times \int \mathrm{d}^4 x\, \mathrm{e}^{\mathrm{i}q.x} \langle p_{li}| j^{\mathrm{EM}}_\mu(x) j^{\mathrm{EM}}_\nu(0) |p_{li}\rangle. \qquad \text{(C.1)}$$

In eqn (C.1) $P(l)$ stands for the probability for the nucleon to go into a specific parton configuration $l$, $f_{li}(\eta)$ is the probability distribution that a parton of type $i$ in configuration $l$ has a four-momentum $p_{li} \simeq \eta p$ (see Chapter 1), and $n_{li}$ is the number of partons of type $i$ (Charge $Q_i$) in the configuration $l$. For spin-$\frac{1}{2}$ partons, the right-hand side of eqn (C.1) is proportional to

$$\int_0^1 \mathrm{d}\eta \sum_l P(l) \sum_i n_{li} Q_i^2 f_{li}(\eta) \,\mathrm{Im\,Tr}\, \eta \not{p} \gamma_\mu S_\mathrm{F}(\eta p + q) \gamma_\nu . \tag{C.2}$$

In the expression of eqn (C.2) we use the result

$$\mathrm{Im}\, S_\mathrm{F}(\eta p + q) \simeq \pi(\eta \not{p} + \not{q}) \delta(q^2 + 2\eta q \,.\, p) \tag{C.3}$$

for leading terms in the limit of large $q^2$. Moreover, we substitute

$$\mathscr{M} \equiv \sum_l P(l) \sum_i n_{li} Q_i^2 f_{li}(w) w \not{p}, \tag{C.4}$$

where $w = -q^2/(2q \,.\, p)$. These operations enable us to write

$$\lim_{\mathrm{Bj}} W^\mathrm{e}_{\mu\nu} \propto \mathrm{Tr}\, \gamma_\mu \frac{\not{q} + w\not{p}}{\nu} \gamma_\nu \mathscr{M}. \tag{C.5}$$

The parton picture for inelastic electron scattering is as in Fig. 42. We introduce the complete set of 16 Dirac matrices $\Gamma^j (j = 1, ---, 16)$ satisfying the Clifford algebra and the completeness condition

$$\sum_j \Gamma^j_{ab} \Gamma_{j,cd} = \delta_{ac} \delta_{bd},$$

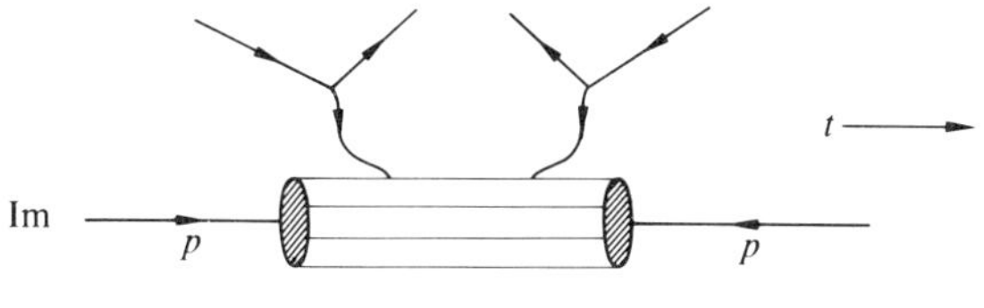


FIG. 42. Inelastic electron scattering according to the parton model ($s$-channel term).

$a, b, c, d$ being spinor indices. Eqn (C.5) may now be rewritten as

$$\lim_{\mathrm{Bj}} W^\mathrm{e}_{\mu\nu} \propto \sum_j \mathrm{Tr}\left(\gamma_\mu \frac{w\not{p} + \not{q}}{\nu} \gamma_\nu \Gamma_j\right) \mathrm{Tr}(\Gamma^j \mathscr{M}). \tag{C.6}$$

Notice that from among the $\Gamma^j$s only $\gamma_\sigma$ and $\gamma_\sigma \gamma_5$ contribute to the right-hand side of eqn. (C.6). This means that

$$\lim_{\mathrm{Bj}} W^\mathrm{e}_{\mu\nu} \propto \frac{(q + wp)_\rho}{\nu} \{\mathrm{Tr}(\gamma_\mu \gamma^\rho \gamma_\nu \gamma^\sigma)\, \mathrm{Tr}(\gamma_\sigma \mathscr{M}) +$$

$$+ \mathrm{Tr}(\gamma_\mu \gamma^\rho \gamma_\nu \gamma^\sigma \gamma_5)\, \mathrm{Tr}(\gamma_\sigma \gamma_5 \mathscr{M})\}. \tag{C.7}$$

Now define

$$J_\sigma(w) \equiv \mathrm{Tr}(\gamma_\sigma \mathcal{M}), \tag{C.8a}$$

$$J_{\sigma 5}(w) \equiv \mathrm{Tr}(\gamma_5 \gamma_\sigma \mathcal{M}). \tag{C.8b}$$

These will be related to the Fourier transforms of the matrix elements of the bilocal currents between nucleon states. Substituting these in eqn (C.7) and making use of the results

$$\tfrac{1}{4}\,\mathrm{Tr}(\gamma_\mu \gamma^\rho \gamma_\nu \gamma^\sigma) = \delta^\rho_\mu \delta^\sigma_\nu + \delta^\sigma_\mu \delta^\rho_\nu - g_{\mu\nu} g^{\rho\sigma} \equiv S_\mu{}^\rho{}_\nu{}^\sigma$$

$$\tfrac{1}{4}\,\mathrm{Tr}(\gamma_\mu \gamma_\rho \gamma_\nu \gamma_\sigma \gamma_5) = \mathrm{i}\varepsilon_{\mu\rho\nu\sigma},$$

we obtain

$$\lim_{\mathrm{Bj}} W^{\mathrm{e}}_{\mu\nu} \propto \frac{1}{\nu}(q+wp)^\rho \{S_{\mu\rho\nu\sigma} J^\sigma(w) + \mathrm{i}\varepsilon_{\mu\nu\rho\sigma} J^\sigma_5(w)\}. \tag{C.9}$$

So far, we have considered only the imaginary part which contains only the direct ($s$-channel) term (see the Introduction, p. 5). For the full amplitude, the ($u$-channel) cross-term, as illustrated in Fig. 43 also contributes. The contribution of this term to $W^{\mathrm{e}}_{\mu\nu}$ vanishes, but adding it all the same, we rewrite eqn (C.9) as

$$\lim_{\mathrm{Bj}} W^{\mathrm{e}}_{\mu\nu} \propto \frac{1}{\nu}(q+wp)^\rho \{S_{\mu\rho\nu\sigma} J^{\ominus,\sigma}(w) + \mathrm{i}\varepsilon_{\mu\nu\rho\sigma} J^{\oplus,\sigma}_5(w)\}, \tag{C.10}$$

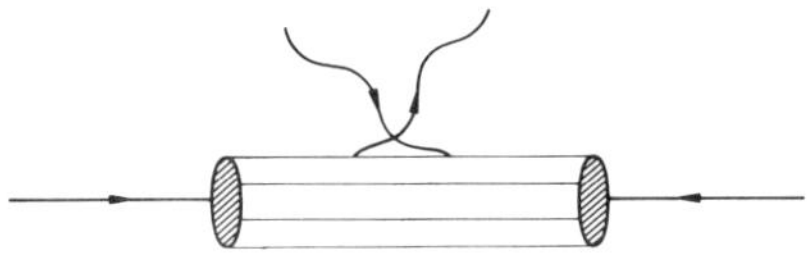

FIG. 43. As Fig. 42 but in the $u$-channel.

where

$$J^{\ominus,\sigma}(w) \equiv J^\sigma(w) - J^\sigma(-w), \tag{C.11a}$$

$$J^{\oplus,\sigma}_5(w) \equiv J^\sigma_5(w) + J^\sigma_5(-w). \tag{C.11b}$$

In an exactly similar manner we may consider $W^5_{\mu\nu}$ involving one axial and another vector current. Then we are led to the result

$$\lim_{\mathrm{Bj}} W^5_{\mu\nu} \propto \sum_j \mathrm{Tr}\left(\gamma_5 \gamma_\mu \frac{\not{q} + w\not{p}}{\nu} \gamma_\nu \Gamma_j\right) \mathrm{Tr}(\Gamma^j \mathcal{M}) +$$

$$+ \text{vanishing cross-term},$$

and finally to

$$\lim_{\mathrm{Bj}} W^5_{\mu\nu} \propto \frac{1}{\nu}(q+wp)^\rho(S_{\mu\rho\nu\sigma}J_5^{\ominus,\sigma}+\mathrm{i}\varepsilon_{\mu\nu\rho\sigma}J^{\oplus,\sigma}). \tag{C.12}$$

Putting in SU(3) matrices also, we easily obtain

$$W^{ab}_{\mu\nu} \propto \frac{(q+wp)^\rho}{\nu}\{\mathrm{i}f^{abc}(S_{\mu\rho\nu\sigma}J^{\oplus,\sigma,c}-\mathrm{i}\varepsilon_{\mu\nu\rho\sigma}J_5^{\ominus,\sigma,c})$$

$$+d^{abc}(S_{\mu\rho\nu\sigma}J^{\ominus,\sigma,c}+\mathrm{i}\varepsilon_{\mu\nu\rho\sigma}J_5^{\oplus,\sigma,c})\}. \tag{C.13}$$

An expression similar to eqn (C.13) obtains for $W^5_{\mu\nu}$. The equivalence with light-cone physics (cf. Chapter 6) is now complete. The explicit occurrence of the light-cone singularities follows from the result

$$\int \mathrm{d}^4q\, \mathrm{e}^{-\mathrm{i}q.x}\frac{1}{\nu}(q+wp)^\rho F(w)$$

$$= -\pi^2\partial^\rho\{\varepsilon(x^0)\delta(x^2)\}f(x\,.\,p),$$

where

$$f(x\,.\,p) = \int_{-\infty}^{\infty} \mathrm{d}\xi\, \mathrm{e}^{\mathrm{i}\xi x.p}F(\xi).$$

The formal statement of parton–light-cone equivalence in the present context is

$$\int \mathrm{d}^4x\, \mathrm{e}^{\mathrm{i}q.x}\varepsilon(x^0)\delta(x^2)\left\langle \mathrm{N}\middle| J^{\ominus,\sigma}\left(\frac{x}{2},-\frac{x}{2}\right)\middle|\mathrm{N}\right\rangle$$

$$\propto \frac{J^{\ominus,\sigma}(w)}{2M\nu} = \frac{1}{2M\nu}\mathrm{Tr}\left\{\gamma^\sigma\sum_l P(l)\sum_i n_{li}Q_i^2 f_{li}(w)w\not{p}\right\}. \tag{C.14}$$

From the above considerations we can clearly see the correspondence between different aspects of the two approaches. The scale invariant algebra of current operators on the light-cone corresponds to the point-like and incoherent behaviour of effectively massless partons, as well as to the symmetry among them. On the other hand, the matrix element of a bilocal current on the light-cone is linked with the momentum distribution of partons. Assumptions about the distribution of partons constituting a physical particle transform into statements about the coordinate space or algebraic structure of the said light-cone matrix element. Moreover, eqn (C.14) shows the parton-light-cone dualism to be a reflection of the wave–particle dualism of quantum mechanics. The former, like the latter, comprises identical physical notions described in momentum space and in configuration space. Nevertheless, there is a fundamental difference between the two dualisms. The one in quantum mechanics is between two supposedly exact pictures. In contrast, both the parton and the light-cone descriptions are approximate, being valid only in certain limited, but not precisely delineated, domains. For deep inelastic lepton–nucleon scattering or in asymptotic $e^+e^-$ annihilation into hadrons (with all the hadronic production modes summed) the two schemes are indeed equivalent. However, when their original frontiers are expanded to include new and more involved reactions (such as semi-inclusive lepton scattering $l + N \rightarrow l' + H +$ 'anything', one-particle inclusive electron positron annihilation $e^+e^- \rightarrow H +$ 'anything', and muon pair production from hadronic collisions $H + N \rightarrow \mu^+\mu^- +$ 'anything' (Jaffe 1972)) in their highly inelastic domains, additional physical inputs need to be grafted onto the two minimal schemes. In these circumstances the strict equivalence between the two descriptions may disappear via the divergence of their domains of validity or otherwise. Broad connections, however, still linger even for these reactions.

Let us now focus our attention on the yet unsolved problems and unclear notions buried in the theoretical ideas described in this book. The central feature that has defied any real understanding so far is the failure of partons (or, more definitively, quarks) to be knocked out in these experiments. If the current is 'truly' a bilinear in quark fields throughout the light-cone, then the cluster decomposition theorem decrees that real quarks must exist. This conclusion may be evaded by using the bilinear structure for only the leading light-cone singularities (or a few non-leading ones in addition) in products of currents. That, however, is begging the question rather than answering it. A more promising approach (Gell-Mann 1972) may be the assignment of para-Fermi statistics of rank 3 (or, equivalently, three colours—cf. Chapter 6) to the quarks; the existence of real (as opposed to virtual) paraparticles may then be forbidden by the minimal extension of one of the postulates of quantum mechanics. Confinement then becomes a consequence of colour. There is already some indirect evidence from the study of Adler–Bell–Jackiw anomalies (Treiman, Gross, and Jackiw 1972) and some support from hadron spectroscopy (Lipkin 1973) for rank-3 paraquarks. Another problem is the presumed fractional charge of the quark and how it manages to redistribute itself among integrally charged hadrons in the final state. This question presents fairly acute difficulties in the quark–parton model, as mentioned in Chapter 2. In the light-cone approach it is merely glossed over by being omitted from the list of quark properties that are abstracted from the underlying field theory. Finally, the consideration of the mass of the quark-parton does lead to a paradox or two. Scale invariance requires quark-partons to be nearly massless; indeed, their masses have been estimated to be in the MeV range (Das, Pandit, and Roy 1973). On the other hand, the calculations of static properties of hadrons in the quark model generally require the quark to be heavier. Some progress has recently been made (Melosh 1974;

Weyers 1973) in understanding the two incarnations of the same object (i.e. the light current quark and the heavy constituent quark) by proposing between them some transformation which may not commute with the quark mass operator.

The development and success of the theoretical ideas of partons, scale invariance, and light-cone physics constitute a renaissance of field-theoretic notions in high-energy physics. Hadrons may be treated as composites of partons whose response to a leptonic probe of large momentum transfer may be predicted from certain Born diagrams, summed incoherently. Alternatively, commutators and products of currents are singular on the light cone; the degrees of the leading light-cone singularities (all $c$-numbers) as well as the asymptotic scale dimension of the associated operators are given canonically by an asymptotically free field theory of massless quarks and gluons. It is an astonishing historical development that, after a prominently undistinguished record in the study of strong interactions, field theory should stage a comeback in deep inelastic lepton–hadron process in a deceptively simple form, i.e. with asymptotic freedom (cf. Chapter 5). Certain hadronic reactions induced by mesons or real photons may be described by matrix elements of commutators of appropriate Bose fields. In such reactions with on-shell particles, the entire light-cone is relevant. Whatever strong-interaction dynamics may control these operators inside the light-cone, it is now an accepted fact that on its surface they have to satisfy the boundary constraints of free or nearly free field theory. For two points with a time-like separation, we may extend the concept of a biolocal current to that of a path-dependent infinitely multi-local current. If $\mathscr{P}$ is a certain path between $x$ and $y$ and $V_\mu$ a vector potential (generalized from the gluon field) arising from strong interactions, we may then have

$$J^{\mathrm{i}}_\mu(x, y, \mathscr{P}) \sim \bar{q}(y)\gamma_\mu \frac{\lambda^{\mathrm{i}}}{2} \exp\left\{\mathrm{i} \oint_{\mathscr{P}\, x}^{y} \mathrm{d}z^\mu \, V_\mu(z)\right\} q(x).$$

It has been suggested (Fritzsch and Gell-Mann 1972) that these may be used to give a 'string' or 'linear rubber-band' formulation (in configuration space) of the dual resonance model. Another exciting possibility is the consistency of the bootstrap scheme of hadrons with the field theory of purely virtual coloured quarks perennially bound because of parastatistics. The optimist will hope that notions such as partons, scale invariance, and light-cone physics will be harbingers of deeper and more far-reaching concepts which can effect such a marriage between field theory and $S$-matrix theory.

# REFERENCES


ADLER, S. L. and DASHEN, R. F. (1968). *Current algebras*, Benjamin, New York.

——— (1974). *Proceedings of the National Accelerator Laboratory Conference on Neutrino Physics* (in press).

BARDEEN, W. A., FRITZSCH, H., and GELL-MANN, M. (1972). *Scale and conformal invariance in particle physics* (ed. R. Gatto). John Wiley and Sons, New York.

BENVENUTTI, A. ET AL. (1974). *Phys. Rev. Lett.* **32**, 125.

BERKELMAN, K. (1971). *Proceedings of the 1971 international symposium on electron and positron interactions at high energies* (ed. N. Mistry). Laboratory of Nuclear Studies, Cornell University.

——— (1972). *Proceedings of the XVI international conference on high energy physics* (ed. J. D. Jackson and A. Roberts), Vol. 4. The National Accelerator Laboratory, U.S.A.

BERMAN, S. M., BJORKEN, J. D., and KOGUT, J. (1971). *Phys. Rev.* **D4**, 3388.

BERNARDINI, C. (1971). *Proceedings of the 1971 international symposium on electron and positron interactions at high energies* (ed. N. Mistry). Laboratory of Nuclear Studies, Cornell University.

BJORKEN, J. D. (1969). *Phys. Rev.* **179**, 1547.

——— (1971*a*) *Proceedings of the Tel Aviv conference on duality and symmetry in hadron physics* (ed. E. Gotsman). Weizmann Science Press, New York.

——— (1971*b*). *Proceedings of the 1971 international symposium on electron and positron interactions at high energies* (ed. N. Mistry). Laboratory of Nuclear Studies, Cornell University.

——— (1973). *Proceedings of the Summer Institute on Particle Physics* (ed. D. W. G. S. Leith and S. D. Drell), Vol. 1. Stanford Linear Accelerator Center.

BJORKEN, J. D. and DRELL, S. D. (1965). *Relativistic Quantum Fields*, McGraw-Hill, New York.

BLOOM, E. D. (1972). *Proceedings of the XVI international conference on high energy physics* (ed. J. D. Jackson and A. Roberts), Vol. 2. The National Accelerator Laboratory, U.S.A.

BONORA L. and VENDRAMIN, I. (1970). *Nuovo Cim.* **70A**, 441.

BRANDT, R. and PREPARATA, G. (1971). *Nucl. Phys.* **B27**, 541.

BROWN, L. S. (1969). *Lectures in theoretical physics* (ed. K. T. Mahanthappa and W. E. Brittin) Vol. XII B. Gordon and Breach, New York.

BROWN, S. G. (1972). *Phys. Rev.* **D5**, 2593.

BURKHARDT, H. and COTTINGHAM, W. N. (1970) *Ann. Phys. (N.Y.)* **56**, 453.

BUDNY, R. (1973). *Nuovo Cim.* **15A**, 173.

———, CHANG, T. H., and CHOUDHURY, D. K. (1972). *Nucl. Phys.* **B44**, 618.

CALLAN, C. (1972*a*). *Phys. Rev.* **D5**, 3202.

——— (1972*b*). *Elementary particle physics: multiparticle aspects* (ed. P. Urban). Springer-Verlag, Berlin.

———, COLEMAN, S., and JACKIW, R. (1970). *Ann. Phys. (N.Y.)* **59**, 42.

CARRUTHERS, P. (1971). *Phys. Rep.* **C1**, No. 1.

CARLSON, C. E. and TUNG, W. K. (1972). *Phys. Rev.* **D5**, 721.

CICCARIELLO, S., GATTO, R., SARTORI, G., and TONIN, M. (1971). *Ann. Phys. (N.Y.)* **65**, 265.

COLEMAN, S. (1971). *Proceedings of the international school of subnuclear physics Ettore Majorana* (ed. A. Zichichi).
——— and JACKIW, R. (1971). *Ann. Phys.* (*N.Y.*) **67**, 552.
——— and GROSS, D. (1973). *Phys. Rev. Lett.* **31**, 851.
CHANG, S. J. and YAN, T. M. (1971). *Phys. Rev.* D**4**, 537.
CHOUDHURY, D. K. (1973). *Nuovo Cim.* **17**A, 189.
CHRIST, N. (1972). *Proceedings of the XVI international conference on high energy physics* (ed. J. D. Jackson and A. Roberts), Vol. 2. The National Accelerator Laboratory, U.S.A.
CORNWALL, J., CORRIGAN, J., and NORTON, R. (1971). *Phys. Rev.* D**3**, 536.
——— and JACKIW, R. (1971). *Phys. Rev.* D**4**, 367.
DAS, T., PANDIT, L. K., and ROY, P. (1973). *Nucl. Phys.* B**53**, 567.
DICUS, D. A., JACKIW, R., and TEPLITZ, V. (1971). *Phys. Rev.* D**4**, 1733.
DONCEL, M. and DE RAFAEL, E. (1971). *Nuovo Cim.* **4**A, 363.
DRELL, S. D. (1969). *Subnuclear phenomena.* International School of Physics 'Ettore Majorana', Course 7 (ed. A. Zichichi). Academic Press, New York.
——— and ZACCHARIASEN, F. (1961). *Electromagnetic structure of the nucleon.* Oxford University Press.
——— and YAN, T. M. (1971). *Ann. Phys.* (*N.Y.*) **66**, 578.
——— and LEE, T. D. (1972). *Phys. Rev.* D**5**, 1738.
DYSON, F. (1966). *Symmetry groups in nuclear and particle physics.* Benjamin, New York.
ELLIS, J. (1970). *Nucl. Phys.* B**22**, 478.
——— (1971). *Nucl. Phys.* B**26**, 537.
——— and JAFFE, R. L. (1973). *Scaling, short distances, and the light-cone.* Notes based on lectures presented at the University of California Santa Cruz Summer School on Particle Physics.
FAYYAZUDDIN and RIAZUDDIN (1972). *Phys. Rev.* **D5**, 2641.
FEYNMAN, R. P. (1969). *High energy collisions.* Third International Conference held at SUNY, Stony Brook (ed. C. N. Yang *et al.*). Gordon and Breach, New York.
——— (1972*a*). *Proceedings of the Neutrino '72 Conference, Balatonfured, Hungary* (ed. A. Frenkel and G. Marx), Vol. II. OMKDK-Technoinform.
——— (1972*b*). *Photon–hadron interactions.* Benjamin, New York.
FRISHMAN, Y. (1971*a*). *Acta Phys. Austriaca* **34**, 351.
——— (1971*b*). *Ann. Phys.* (*N.Y.*) **66**, 373.
——— (1972). *Proceedings of the XVI international conference on high energy physics* (ed. J. D. Jackson and A. Roberts), Vol. 4. The National Accelerator Laboratory, U.S.A.
FRITZSCH, H. and GELL-MANN, M. (1971*a*). *Broken scale invariance and the light cone* (ed. M. Gell-Mann and K. Wilson) Gordon and Breach, New York.
——— and GELL-MANN, M. (1971*b*). *Proceedings of the Tel Aviv conference on duality and symmetry in hadron physics* (ed. E. Gotsmann). Weizmann Science Press, New York.
——— and GELL-MANN, M. (1972). *Proceedings of the XVI international conference on high energy physics* (ed. J. D. Jackson and A. Roberts), Vol. 2. The National Accelerator Laboratory, U.S.A.
FUBINI, S. and FURLAN, G. (1968). *Ann. Phys.* (*N.Y.*) **48**, 322.
GASIOROWICZ, S. (1966). *Elementary particle physics.* John Wiley & Sons, New York.
GELL-MANN, M. (1969). *Proceedings of the third Hawaii topical conference on particle physics.* Western Periodicals, Los Angeles.

——— (1972). *Elementary particle physics: multiparticle aspects* (ed. P. Urban). Springer-Verlag, Berlin.

——— and NEE'MAN, Y. (1964). *The eightfold way*. Benjamin, New York.

GRIBOV, V. N. (1968). *Soviet Phys. JETP* **26**, 414.

GROSS, D. and WILCZEK, F. (1973). *Phys. Rev.* **D8**, 3633; *ibid* **D9**, 980 (1974).

——— and TREIMAN, S. B. (1971). *Phys. Rev.* **D4**, 1059.

GUNION, J. F., BRODSKY, S. J., and BLANKENBECLER R. (1972). *Phys. Rev.* **D6**, 649; (1973) *ibid.* **D8**, 287.

HEY, A. J. G. and MANDULA, J. (1972). *Phys. Rev.* D5, 2610.

HOFSTADTER, R. (1963). *Electron scattering and nuclear and nucleon structure*, Benjamin, New York.

HUANG, K. (1972). *Proceedings of the 14th Latin American School of Physics, Caracas, Venezuela* (in press).

JACKIW, R. (1971*a*). *Proceedings of the international school of subnuclear physics 'Ettore Majorana'* (ed. A. Zichichi).

——— (1971*b*). *Phys. Rev.* D3, 2005.

——— and WALTZ, R. E. (1972). *Phys. Rev.* D6, 702.

——— (1972*a*). *Springer Tracts Mod. Phys.* **62**, 1.

——— (1972*b*). *Physics to-day* January, p. 23.

——— (1972*c*). *Comm. Nucl. Particle Phys.* **5**, 153.

JARLSKOG, C. and ROY, P. (1972). *Nucl. Phys.* **B48**, 415.

JAFFE, R. (1972*a*). *Phys. Rev.* **D5**, 2622.

——— (1972*b*). *Phys. Rev.* **D6**, 716.

KENDALL, H. (1971). *Proceedings of the 1971 international symposium on electron and positron interactions at high energies* (ed. N. Mistry). Laboratory of Nuclear Studies, Cornell University.

KOGERLER, R. (1972). *Introductory deep inelastics*. Lectures delivered at the Chalmers Tekniska Hogskola, Goteborg.

KOGUT, J. and SOPER, D. (1970). *Phys. Rev.* **D1**, 290.

——— and SUSSKIND, D. (1973). *Phys. Rep.* **8C**, No. 2.

KOKKEDEE, J. J. J. (1969). *The quark model*. Benjamin, New York.

KUTI, J. and WEISSKOPF, V. (1971). *Phys. Rev.* **D4**, 3418.

LANDSHOFF, P. V. and POLKINGHORNE, J. C. (1972). *Phys. Rep.* **5C**, No. 1.

LEE, T. D. (1972). *Phys. Rev.* **D6**, 1120.

LEUTWYLER, H. (1969). *Springer tracts Mod. Phys.* **50**, 29.

——— (1972*a*). *Strong interaction physics, Lecture notes in physics* (ed. W. Ruhl and A. Vancura), Vol. 17.

——— (1972*b*). *Ann. Phys. (N.Y.)* **74**, 524.

——— and OTTERSON, P. (1972). *Scale and conformal invariance in particle physics* (ed. R. Gatto). John Wiley and Sons, New York.

LIGHTHILL, M. J. (1964). *Introduction to Fourier analysis and generalized functions.* Cambridge University Press.

LIPKIN, H. (1972). *Phys. Rev. Lett.* **28**, 63.

——— (1973). *Phys. Rep.* **8C**, No. 3.

LITTLE, R. (1972). Footnote 76 of Frishman (1972).

LLEWELLYN-SMITH, C. H. (1971). *Phys. Rev.* **D4**, 2392.

——— (1972*a*). *Springer Tracts mod. Phys.* (ed. G. Hohler). **62**, 51.

——— (1972*b*). *Proceedings of the fourth international conference on high energy collisions* (ed. J. R. Smith). Oxford University Press.
——— (1972*c*). *Phys. Rep.* **3**C, No. 5.
MACK, G. (1967). *Partially conserved dilation current*. Ph.D. Thesis, University of Berne, Switzerland.
——— (1971). *Nucl. Phys.* B**35**, 592.
MAJUMDAR, D. P. (1971). *Phys. Rev.* D**3**, 2869.
MARSHAK, R, RIAZUDDIN, and RYAN, C. (1969). *Theory of weak interactions in particle physics*. Interscience, New York.
MARTIN, P. (1967). *Proceedings of the international conference on electron and photon interactions at high energies*. SLAC, Stanford University.
MELOSH, H. (1974). *Phys. Rev.* D**9**, 1075.
MEYER, P. (1972). *Proceedings of the Cargese Summer Institute* (in press).
MILLER, G. *et al* (1972). *Phys. Rev.* D**5**, 528.
NACHTMANN, O. (1971). *J. Phys.* (*Fr.*) **32**, 97.
——— (1972*a*). *Nucl. Phys.* B**38**, 397.
——— (1972*b*). *Phys. Rev.* D**5**, 686.
NASH, C. (1972). *Nuovo Cim.* **8**A, 300.
PANDIT, L. K. (1973). *Advances in High Energy Physics*, (ed. S. M. Roy and V. Singh) Vol. 1. Tata Institute of Fundamental Research, Bombay, India.
PERKINS, D. H. (1972). *Proceedings of the XVI international conference on high energy physics* (ed. J. D. Jackson and A. Roberts), Vol. 4. The National Accelerator Laboratory, U.S.A.
POLKINGHORNE, J. C. (1972). *Nuovo Cim.* **8**A, 592.
POLYAKOV, A. (1971). *Proceedings of the international school of physics*, Erevan (in press).
ROY, P. (1972). *Phys. Rev.* D**5**, 1180.
——— (1972). *Proceedings of the first symposium on high energy physics* (ed. P. K. Malhotra, *et al.*) Indian Institute of Technology, Bombay.
SAKURAI, J. J. (1967). *Advanced quantum mechanics*. Addison-Wesley, New York.
SCHWEBER, S. S. (1962). *An introduction to quantum field theory*. Harper and Row, New York.
SCHROER, B. (1972). *Proceedings of the IV simposio Brasileiro de fisica teorica*, Rio de Janeiro, January 1972 (in press).
SILVESTRINI, V. (1972). *Proceedings of the XVI international conference on high energy physics* (ed. J. D. Jackson and A. Roberts), Vol. 4. The National Accelerator Laboratory, U.S.A.
SOGARD, M. R. (1972). *Proceedings of the XVI international conference on high energy physics* (ed. J. D. Jackson and A. Roberts) Vol. 4. The National Accelerator Laboratory, U.S.A.
TAYLOR, R. (1969). *Proceedings of the 4th international symposium on electron and photon interactions at high energies* (ed. D. W. Braben). Daresbury Nuclear Physics Laboratory, U.K.
TREIMAN, S. B., GROSS, D., and JACKIW, R. (1972). *Lectures on current algebra*. Benjamin, New York.
WEINBERG, S. (1966). *Phys. Rev.* **150**, 1313.
WEYERS, J. (1973). *Proceedings of the International Summer School on particle interactions at very high energies, Louvain* (in press).

WILSON, K. (1969). *Phys. Rev.* **179**, 1499.
——— (1970*a*). *Proceedings of the Midwest conference on theoretical physics.* University of Notre Dame.
——— (1970*b*). *Phys. Rev.* D**2**, 1473; (1970*c*) *ibid.* D**2**, 1478.
——— (1971*a*). *Proceedings of the 1971 international symposium on electron and positron interactions at high energies* (ed. N. Mistry). Laboratory of Nuclear Studies, Cornell University.
——— (1971*b*). *Phys. Rev.* D**3**, 1818.
——— (1972). *Phys. Rev.* D**6**, 419.
——— and ZIMMERMANN, W. (1972). *Commun. math. Phys.* **24**, 87.
WRAY, D. (1972). *Nuovo Cim.* **9**A, 463.
YAN, T. M. (1973). *Recent advances in particle physics* (ed. F. Cooper), Annals of the New York Academy of Sciences, Vol. 229.
ZEE, A. (1973). *Phys. Rev.* D**7**, 3540.
ZIMMERMANN, W. (1971). *Proceedings of the symposium on basic questions in elementary particles physics*, Max Plank Institut, München, Germany.
ZUMINO, B. (1970). *Lectures on elementary particles and quantum field theory, Brandeis Summer School* (ed. S. Deser *et al.*) M.I.T. Press.

# SUBJECT INDEX